AF678561

Trajectories of Chemistry Education Innovation and Reform

ACS SYMPOSIUM SERIES **1145**

Trajectories of Chemistry Education Innovation and Reform

Thomas Holme, Editor
Iowa State University
Ames, Iowa

Melanie M. Cooper, Editor
Michigan State University
East Lansing, Michigan

Pratibha Varma-Nelson, Editor
Indiana University-Purdue University Indianapolis
Indianapolis, Indiana

Sponsored by the
ACS Division of Chemical Education

American Chemical Society, Washington, DC

Distributed in print by Oxford University Press

Library of Congress Cataloging-in-Publication Data

Trajectories of chemistry education innovation and reform / Thomas Holme, editor, Iowa State University, Ames, Iowa, Melanie M. Cooper, editor, Michigan State University, East Lansing, Michigan, Pratibha Varma-Nelson, editor, Indiana University-Purdue University Indianapolis, Indianapolis, Indiana ; sponsored by the ACS Division of Chemical Education.

pages cm. -- (ACS symposium series ; 1145)

Includes bibliographical references and index.

ISBN 978-0-8412-2888-7 (alk. paper)

1. Chemistry--Study and teaching--Congresses. I. Holme, Thomas A., editor of compilation. II. Cooper, Melanie M., editor of compilation. III. Varma-Nelson, Pratibha, editor of compilation. IV. American Chemical Society. Division of Chemical Education, sponsoring body.

QD40.T65 2013

540.71--dc23

2013035969

The paper used in this publication meets the minimum requirements of American National Standard for Information Sciences—Permanence of Paper for Printed Library Materials, ANSI Z39.48n1984.

Copyright © 2013 American Chemical Society

Distributed in print by Oxford University Press

All Rights Reserved. Reprographic copying beyond that permitted by Sections 107 or 108 of the U.S. Copyright Act is allowed for internal use only, provided that a per-chapter fee of $40.25 plus $0.75 per page is paid to the Copyright Clearance Center, Inc., 222 Rosewood Drive, Danvers, MA 01923, USA. Republication or reproduction for sale of pages in this book is permitted only under license from ACS. Direct these and other permission requests to ACS Copyright Office, Publications Division, 1155 16th Street, N.W., Washington, DC 20036.

The citation of trade names and/or names of manufacturers in this publication is not to be construed as an endorsement or as approval by ACS of the commercial products or services referenced herein; nor should the mere reference herein to any drawing, specification, chemical process, or other data be regarded as a license or as a conveyance of any right or permission to the holder, reader, or any other person or corporation, to manufacture, reproduce, use, or sell any patented invention or copyrighted work that may in any way be related thereto. Registered names, trademarks, etc., used in this publication, even without specific indication thereof, are not to be considered unprotected by law.

PRINTED IN THE UNITED STATES OF AMERICA

Foreword

The ACS Symposium Series was first published in 1974 to provide a mechanism for publishing symposia quickly in book form. The purpose of the series is to publish timely, comprehensive books developed from the ACS sponsored symposia based on current scientific research. Occasionally, books are developed from symposia sponsored by other organizations when the topic is of keen interest to the chemistry audience.

Before agreeing to publish a book, the proposed table of contents is reviewed for appropriate and comprehensive coverage and for interest to the audience. Some papers may be excluded to better focus the book; others may be added to provide comprehensiveness. When appropriate, overview or introductory chapters are added. Drafts of chapters are peer-reviewed prior to final acceptance or rejection, and manuscripts are prepared in camera-ready format.

As a rule, only original research papers and original review papers are included in the volumes. Verbatim reproductions of previous published papers are not accepted.

ACS Books Department

Contents

Introductory Chapters

Trajectories of Reform in Curricula

Trajectories of Reform of Teaching Tools and Methods

Trajectories of Reform of Institutions and Institutional Collaboration

Indexes

Introductory Chapters

Chapter 1

Importance of Considering Longitudinal Trajectories in Education Reform Efforts

Thomas A. Holme,[*,1] Melanie M. Cooper,[2] and Pratibha Varma-Nelson[3]

[1]Department of Chemistry, Iowa State University, Iowa State University, 0213 Gilman Hall , Ames, Iowa 50011
[2]Department of Chemistry, Michigan State University, 578 S Shaw Lane, East Lansing, Michigan 48824
[3]Department of Chemistry and Chemical Biology and Center for Teaching and Learning, Indiana University-Purdue University, 755 W. Michigan Street, Indianapolis, Indiana 46202
[*]E-mail: taholme@iastate.edu

This chapter introduces the collection of articles in this book with an emphasis on why it is important to consider the way that educational research and reform efforts change over time. The importance of considering a longitudinal view of education reform is emphasized in two ways. First, the context of this work relative to current literature is considered. Second, the idea of a greater focus on the longer-term trajectories of reform efforts in considered in terms of suggestions for the future of chemistry education.

Introduction

The Symposium Series of books from the American Chemical Society (ACS) serves as a repository of important trends in chemical science and education. This collection provides, in essence, a set of snapshots of the field and helps establish matters of sufficient importance to merit discussion, by highlighting the topics of specific symposia held at ACS scientific meetings. This particular volume fits within this paradigm well. It arises from a symposium held to acknowledge and celebrate the efforts of Dr. Susan Hixson as a program officer in the Division of Undergraduate Education at the National Science Foundation (NSF), on the

© 2013 American Chemical Society

occasion of her retirement from this position. Funding for projects in science or science education has inherent importance for any of a variety of reasons, but this symposium was not rooted in the economics, but rather in the sense of the continuity of leadership throughout an array of changes in how reform was approached by the NSF. In a practical sense, what the continuity of the permanent program officers provides is a means by which reform efforts can grow incrementally, even while specific funding initiatives come and go. This symposium, therefore, provided a moment to look at the trajectories of reform, and it served as the generating moment for this volume.

The broad concept of educational reform in science and particularly within chemistry is a pervasive one in the United States and has been for decades (*1–4*). Nonetheless, the ability to enact large scale change, based on theories and evidence of efficacy has been modest at best. This collection of articles offers the suggestion that the fragmented nature of many reform efforts represents one critical reason for the modest success. By gathering a group of articles that describe reform endeavors that have been sustained over some length of time, we have sought to start to exemplify the importance of continuity in funding for both reform efforts and the concomitant assessment of the outcomes of these reforms.

Beyond the evidence associated with the existence of this collection of articles, it is also possible to consider the concept of trajectories of reform efforts within the context of understanding how either science or education change. We will describe several such ways to consider this body of work in the next section and then highlight the connection of the articles to each other and to this literature. Finally, we will summarize our impressions of the possible mechanisms for moving from the points on the trajectories noted here to the future.

Models and Theories of Change in Science and Education

The confluence of educational practice and science practice as it emerges in college chemistry courses plays an important role in understanding what changes may be possible in the teaching of chemistry. Studies associated with change in higher education can often identify structural factors within academia that serve to limit the prospects for reform (*5–8*), and there is little evidence that single studies disseminating new curriculum or practices have a major impact on practice (*9*). This collection of studies is almost unique in that it takes a historical view of change, over the past twenty years or so, and provides evidence for how change might be accomplished, at various grain sizes, and in a range of settings. Even so, these large grain views of higher education tend to not account explicitly for specific characteristics of particular disciplines, in this case chemistry. Even when compared with other science disciplines, the classroom practice of chemistry is subtly different (*10*). As noted in a recent report on Discipline Based Education Research (DBER) (*11*), these differences accentuate the motivation for educational research being conducted within the confines of specific disciplines.

While the DBER report concludes that the different areas of DBER are loosely connected disciplines with closer ties to their parent disciplines than to each other, the conclusions and recommendations are all applicable to chemistry.

For example, it is widely documented that many college students hold incorrect beliefs that are difficult to "overcome", particularly for concepts that involve very small or very large spatial or temporal scales. Clearly, this concern is particularly problematic for chemistry, since a robust understanding of molecular level interactions and processes is necessary. This, coupled with another finding from the DBER report, that serious impediments to learning emerge from difficulties with disciplinary specific representations such as chemical structures, means that there are specific difficulties in chemistry that instructors and curriculum developers must be aware of. The report suggests that these difficulties require integrating proven strategies for general instruction (such as socially mediated learning) with targeted instruction aimed at helping students overcome these specific challenges to learning.

The DBER report also suggests that future studies on to best facilitate the translation of DBER into practice. The extent of education research dissemination requires more nuanced, multi faceted investigations than are currently available, but as of now there is little evidence of widespread adoption of evidence-based approaches to teaching and learning at the college level. However, productive change is more likely if efforts are "1) consistent with research on motivating adult learners, 2) include a deliberate focus on changing faculty conceptions about teaching and learning, 3) recognize the cultural and organizational norms of the department and institution, and 4) work to address those norms that pose barriers to change in teaching practice (*11*)."

One way to consider the nature of chemistry education reform efforts is to view the potential barriers to change as contradicting claims on educational resources (*12*). In principle, with infinite, or much larger resources, the barriers to change would be less – perhaps even minimal. When cast in this light, a theme that emerges in looking at reform efforts over longer time-scales is that halting change stems from the time it takes to make sense out of conflicting data. A key example of this type of challenge arises fairly often, when measures of student learning, particularly content tests, do not show large gains after a teaching innovation has been implemented. One possible explanation for this observation is that such tests do not necessarily measure what the innovation was meant to promote. Without sustained research, however, it is difficult to definitively know the cause.

Another aspect of educational change that merits consideration is the cultural background in which it occurs. Considerable efforts have been made over the years to understand the nature of cultural capital in science education (*13*). For example, it has been argued (*14*) that for most students, the science classroom represents a sub-culture that is quite distinct from their daily experience (with family or peers, for instance) and one result is that many students routinely compartmentalize the science knowledge (*15–17*). The challenge of simultaneously supporting content-based strategies for education reform with other cognitive strategies or socio-cultural strategies remains an important one to consider. Arguably, the only way these aspects can be considered is with longer-term work as represented in the idea of trajectories in this volume.

Another confounding component of educational reform efforts lies in the nature of replicated studies (*18*). The premise that replication of the results of an educational research study in a new context will lead to a new, or improved, understanding of student learning in either context is not always obvious. Identical results in different contexts, for example, would seem rather suspicious, but if learning gains for students are lower in the new context is the value of the original research lessened? This type of question clearly cannot be addressed by single instance education reform efforts. Questions such as these argue forcefully the importance of considering trajectories that reform efforts acquire as they move forward, and this volume accentuates several such instances within chemistry education reform.

Summary of Studies in This Volume

The studies presented in this volume are organized into four sections. The introductory section includes this paper and an additional paper authored by Hixson titled, "Trends in NSF-Supported Undergraduate Chemistry Education, 1992-2012" (Chapter 2). This paper connects strongly to the motivation of the ACS Symposium that represents the origin of this project because it summarizes the grant funding trajectory of the National Science Foundation as it related to chemistry education for the past 20 years.

The next section of the volume includes four papers that are generally related to the trajectory taken to accomplish curricular reform efforts. In part because the number of students involved in the course, these papers reflect the relatively high concentration of work at the General Chemistry level. The first paper, "Research on Learning in the Chemistry Laboratory: A Trajectory Connecting Student Outcomes to Thinking Processes" (Chapter 3) by Rickey and Tien, describes the development and impact of a teaching strategy called MORE (Model – Observe – Reflect – Explain) that employs guided discovery methods to improve student understanding and retention of chemistry concepts. The next paper is "Twenty Years of Learning in the Cooperative General Chemistry Laboratory" (Chapter 4) by Cooper and Sandi-Urena. This paper provides a historical account of a reform of general chemistry labs at one institution and the research efforts that emerged over the years, as the authors developed expertise and an understanding of how laboratory activities might affect outcomes for both the students and the graduate teaching assistants.

The third paper in this section describes a number of strategies, in terms of content and in terms of teaching strategies, that were used to reform a specific course over time. The paper "A Trajectory of Reform in General Chemistry for Engineering Students" (Chapter 5) by Holme and Caruthers has a focus on the idea that service courses like General Chemistry have constraints and opportunities associated with the student clientele of the course. The final paper in this section, "Developing a Content Map and Alignment Process for the Undergraduate Curriculum in Chemistry" (Chapter 6) by Zenisky and Murphy, includes significant information about general chemistry, but also extends to the rest of the undergraduate chemistry major. This paper emphasizes a way to vet efforts in chemistry among many stakeholders, essentially establishing a trajectory within a broader community of educational researchers and practitioners over time.

The third section of papers advances the theme of considering reform in chemistry education by enhancing teaching methods and tools available for the effort. The first paper in this section, "PLTL: Tracking the Trajectory from Face-to-Face to Online Environments" (Chapter 7) by Varma-Nelson and Banks, emphasizes a specific segment of a trajectory in the use of Peer Led Team Learning (PLTL). In this case, the emphasis is on porting a successful innovation in the traditional classroom environment and describing the trajectory that allows this method to move to a new, electronic format. The second paper in this section, "Working To Build a Chemical Education Practice" (Chapter 8) by Wink, Fetzer Gisalson, and Ellefson, emphasizes how different settings for educational reform can nonetheless lead to commonalities in the development of both teachers and curricula over time.

The third paper in this section, "The Evolution of Calibrated Peer Review" (Chapter 9) by Russell, follows a long-term development of a specific teaching tool (CPR) that allows instructors to incorporate writing into even large courses. The ways in which this tool developed and how the developers changed the system in response to an expanding user base are key themes of this chapter. The fourth article in this section, "A Chronology of Assessment in Chemistry Education" (Chapter 10) by Bretz, takes a long-term view of how curricular reform efforts collect and make sense out of data about efficacy. The acceleration of the role of assessment during the past 20 years of reform efforts represents an important aspect of this topic.

The final section of the volume emphasizes the role of institution-wide reform efforts and the importance of reform over multiple institutions. The first paper in this section, "Lessons Learned from Collaborations in Chemistry Assessment across Universities: Challenges in Transfer and Scale" (Chapter 11) by Paek and Holme, looks at a specific collaborative effort to leverage several individual projects into a larger vehicle for change. The emphasis of this project on assessment meshes with the final chapter of the previous section. The second paper in this section, "Undergraduate Research with Community College Students: Models and Impacts" (Chapter 12) by Higgins, focuses on two key aspects of student learning. First, the power of undergraduate research is emphasized. Second, the role of two-year colleges is also a key factor in the projects described here. With the large number of students who take chemistry in these schools, this emphasis is particularly important.

The third paper in this section, "Preparing the Future STEM Faculty: The Center for the Integration of Research, Teaching, and Learning" (Chapter 13) by Mathieu, takes essentially a dual-trajectory approach. The first trajectory describes how a multiple-institution effort can be initiated and sustained. The second trajectory is that long-term sustainability of education reform efforts depends strongly on teaching the future professoriate, and this project works directly in this area. The fourth and final chapter in this section and in the book is "Improving STEM Student Success and Beyond: One STEP at a Time" (Chapter 14) by Scharburg. This chapter takes a long-term look at how reform efforts that are initiated in a single department or college within an institution can spread over time and improve student outcomes in a wider array of programs at that school.

Moving Forward in Chemistry Education Reform

What is clear from this collection of reports on trajectories to reform is that reform is possible, but that it takes time, resources, and an awareness of the specific difficulties that chemistry learner's face. At the same time, it is important to be aware that there are large gaps in the research. As yet, we know very little about how specific reform efforts affect different populations of students. For example, few studies are disaggregated by sex, socioeconomic background, race/ethnicity, or ability. We know little about how students at different stages of their academic careers are affected by changes, or how difficulties first identified in introductory courses "play out" as students move through a curriculum. Are innovations designed to have an impact on students in one learning environment effective for students in a different environment? Some studies have found that teaching strategies or methods show improved outcomes in various environments (*19*), but even then, more studies are needed to identify factors that encourage successful cross-cutting effects and those that hinder such effects. We have few longitudinal studies that investigate how change affects a learning environment and outcomes over time, and how these changes affect retention in STEM disciplines. Despite current enthusiasms for online learning, we do not have convincing studies on the differences between face-to-face environments and online environments, and what that means for chemistry education reform. More generally, we need more studies about reform in general. What is the role of the reward system? How can we change institutional and departmental cultures so that evidence based teaching and learning becomes the norm?

Our assessment methods and techniques must improve and address new outcomes, as we begin to understand that learning chemistry means more than chemistry disciplinary knowledge, but also includes the development of science practices such as the use and construction of models, and the development of explanations and arguments. If we believe that there is more to learning in the laboratory, for instance, than replicating exercises and confirming data, then we must focus on developing ways to assess the outcomes we value.

If the collection of articles in this volume tells us anything, it is that addressing key questions such as these will take time and concerted efforts. The challenge of sustaining reform can only be met by affording the time to consider the needs and interests of a range of stakeholders. Thus, taking time to take stock of trajectories of reform represents a crucial exercise in shaping meaningful reform efforts for the future. We hope this collection of articles provides an example of why this introspection is worthwhile. There is little doubt that the over-arching goal of all the projects noted here, improving the ability of students to learn chemistry, is vital for any number of reasons. This fact makes it worth following reform efforts over time and characterizing the paths taken and the lessons learned.

References

1. Lagowski, J. J. The continuing attempt at reform in science education. *J. Chem. Educ.* **1988**, *65*, 1.
2. Lloyd, B. W.; Spencer, J. N. The forum: New directions for general chemistry: Recommendations of the task force on the general chemistry curriculum. *J. Chem. Educ.* **1994**, *71*, 206–209.
3. Fahmy, A. F. M.; Lagowski, J. J. Systemic reform in chemical education: An international perspective. *J. Chem. Educ.* **2003**, *80*, 1078–1083.
4. Cooper, M. M. The case for reform of the undergraduate general chemistry curriculum. *J. Chem. Educ.* **2010**, *87*, 231–232.
5. Kezar, A. *Understanding and Facilitating Organizational Change in the 21st Century: Recent Research and Conceptualizations*; ASHE-ERIC Higher Education Reports; Washington, DC, 2001.
6. Woodbury, S.; Gess-Newsome, J. Overcoming the paradox of change without difference: A model of change in the arena of fundamental school reform. *Educ. Policy* **2002**, *16*, 763–792.
7. Henderson, C.; Dancy, M. H. Barriers to the use of research-based instructional strategies: The influence of both individual and situational characteristics. *Phys. Rev. Special Topics – Phys. Educ. Res.* **2007**, *3* (020102), 1–14.
8. Hearne, J. L.; Henkin, A. B.; Dee, J. R. Enabling initiative and enterprise: Faculty-led course redesign in a STEM discipline. *Educ. Res. Q.* **2011**, *35*, 33–62.
9. Henderson, C.; Beach, A.; Finkelstein, N. Facilitating change in undergraduate STEM instructional practices: An analytic review of the literature. *J. Res. Sci. Teach.* **2011**, *48*, 952–984.
10. Hora, M. T.; Jerrare, J. J. Instructional systems of practice: A multidimensional analysis of math and science undergraduate course planning and classroom teaching. *J. Learn. Sci.* **2013**, *22*, 212–257.
11. National Research Council. *Discipline Based Education Research: Understanding and Improving Learning in Undergraduate Science and Engineering*; National Academies Press: Washington, DC, 2012
12. Kahveci, A.; Gilmer, P. J.; Sotherland, S. A. Understanding chemistry professors' use of educational technologies: An activity theoretical approach. *Int. J. Sci. Educ.* **2008**, *30*, 325–351.
13. Claussen, S.; Osborne, J. Bourdieu's notion of cultural capital and its implications for the science curriculum. *Sci. Educ.* **2013**, *97*, 58–79.
14. Aikenhead, G. S. Science education: Border crossing into science education. *Stud. Sci. Educ.* **1996**, *27*, 1–52.
15. Tiechert, M. A.; Tien, L. T.; Anthony, S.; Rickey, D. Effects of context on students' molecular-level ideas. *Int. J. Sci. Educ.* **2008**, *30*, 1095–1114.
16. Hansson, L.; Lindahl, B. I have chosen another way of thinking. *Sci. Educ.* **2010**, *19*, 895–918.
17. Cooper, M. M.; Underwood, S. M.; Hilley, C. Z.; Klymkowsky, M. W. Development and assessment of a molecular structure and properties learning Progression. *J. Chem. Educ.* **2012**, *89*, 1351–1357.

18. Taber, K. S. Vive la différence? Comparing "like with like" in studies of learners' ideas in diverse educational contexts. *Educ. Res. International* **2012**, DOI:10.1155/2012/168741.
19. Gafney, L.; Varma-Nelson, P. *Innovations in Science Education and Technology: Peer-Led Team Learning: Evaluation, Dissemination, and Institutionalization of a College Level Initiative*; Springer, Weston, MA, 2008; p 16.

Chapter 2

Trends in NSF-Supported Undergraduate Chemistry Education, 1992-2012

Susan H. Hixson*

Arlington, Virginia 22207, United States
***E-mail: hixsonsusan@gmail.com**

During the period 1992-2012 the Division of Undergraduate Education at the National Science Foundation catalyzed significant changes in the undergraduate educational efforts by faculty in the chemistry community through major expansions of the programs that it supported. Over the two decades, the disciplinary programs supported by the division evolved from an early emphasis on instrumentation and content to a broader focus that included content, pedagogy, cyber applications, assessment, education research, and evaluation. The Chemistry Initiative, a systemic effort launched in 1994, proved to be a significant factor in expanding and connecting the faculty in the undergraduate chemistry education community. In addition, during the period 1992-2012 a number of broader-based programs were established that led chemists to become involved in efforts to prepare future K-12 teachers, promote improvement in advanced technological education at community colleges, create and collect quality education materials in online collections, and increase the number of undergraduates receiving science, technology, engineering and mathematics (STEM) degrees. A unique large-scale project strengthened traditional graduate student training at major research institutions by encouraging these future STEM faculty to receive significant experience with educational issues.

© 2013 American Chemical Society

Introduction

The period 1992-2012 was chosen for this review because this timeframe coincides with the years during which the Division of Undergraduate Education (DUE) at the National Science Foundation (NSF) catalyzed significant changes in the undergraduate educational efforts by faculty in the science, technology, engineering, and mathematics (STEM) community through major expansions of the programs that it supported (*1*).

Role of DUE

The prime location for leadership and support of undergraduate chemistry education at the NSF is DUE within the Directorate for Education and Human Resources (EHR). The sole responsibility of DUE has been the support of undergraduate education across all STEM fields, and DUE has provided the leadership for the NSF in the development of programs that encourage improvement in the national undergraduate STEM academic environment.

The EHR Directorate was abolished by the Reagan administration in 1981. The decision was reversed in 1983, and the first restored support was primarily for K-12 education. By 1987 funding for undergraduate education was represented only by a $7.5-million program for support of college instrumentation (*2*). However, by the mid-1990's an array of programs was in existence in DUE, and these programs evolved and expanded, with additional programs added over the next two decades.

Role of the Division of Chemistry

A second location at NSF provides leadership and support for several more focused efforts in undergraduate chemistry education. In the Directorate for Mathematical and Physical Sciences, the Division of Chemistry has as its prime focus the support of cutting-edge research. In addition, the Division manages the Research Experiences for Undergraduates (NSF 12-569 (*3*)) program that supports research by undergraduate students, typically through summer experiences. The Division remained a staunch supporter of this undergraduate effort throughout the period 1992-2012. The program evolved from the traditional support in the early 1990's of upper-level chemistry majors to support that encourages new options in order to broaden the pool of undergraduate students that undertake research in chemistry. By 2012 the program included support for lower-level undergraduate students and community college students and for projects that provide international research experiences. The Division also provided opportunities throughout 1992-2012 for institutions to obtain instrumentation that is required for undergraduate research programs at colleges and universities. Finally, a number of short-term, experimental programs for undergraduate education were offered. For example, under the leadership of Division Director Arthur Ellis (2002-2006), the Division ran several competitions of the Undergraduate Research Centers/Collaboratives program (NSF 03-595, NSF 05-539, NSF 06-521) that called for projects up to $2.7 million in size for developing new, multi-institutional undergraduate research models that involved significant numbers of first- and second-year students at a lower cost per student than the traditional "individual students working in laboratories" model. The chapter by Thomas Higgins (*4*) in this Book provides an example of a project supported through this program.

Critical Factors in the Evolution of DUE

Three factors in DUE were critical for the evolution of the support for undergraduate STEM education over the years 1992-2012. First, two remarkable Division Directors provided thoughtful and consistent leadership for DUE. These two Directors also established the system within DUE in which program officers worked within their disciplines in managing some programs, and worked as multidisciplinary teams in managing other programs. This system gave DUE the flexibility to run a broad array of programs with a limited staff, allowing each program to be managed by program officers who brought the necessary mix of discipline-specific expertise to each program.

Robert F. Watson, a chemist, served at the NSF in science education in the pre-Reagan years, and continued to serve after the restoration of EHR through 1996 as the first Division Director of DUE and its predecessor division, the Division of Undergraduate Science, Engineering, and Mathematics Education. Norman L. Fortenberry, an engineer, continued the development within DUE from 1996-2002 as Division Director. Each of these Directors had very broad and deep knowledge of the undergraduate STEM community, so they were able to bring faculty and organizations together to initiate new efforts and to extend on-going efforts across disciplines or across institutions. Each of these Directors was heavily involved in program development, joining their own expertise together with the expertise of the program officers in the division. Finally, each of these Directors was respected across NSF and was able to develop efforts that were relevant to the interests of the research directorates.

The second factor that was critical in the evolution of DUE was the support of outside groups that brought their concerns to Congress. Most funding for specific programs at NSF is requested by NSF from Congress and represents priorities developed by NSF. However, the majority of DUE's funding over the period 1992-2012 came through programs initiated and mandated by Congress. Both Dr. Watson and Dr. Fortenberry played significant roles in working with Congressional staff to shape these programs. Thus, most of these Congressionally-mandated programs served the mission of DUE well.

The third factor that was critical in the evolution of DUE was the presence of the program officers who served in the division. These faculty brought to bear on their work in DUE their STEM disciplinary expertise, their faculty and administrative experiences gained at their home institutions, and their personal professional networks. Table I shows the faculty who served as chemists in DUE from the time that the division was reinstituted. Almost all of these program officers served for one or two years as rotators in the division, returning to their home institutions or moving to a position at another institution. They generally went on to serve as leaders in national undergraduate efforts after their stints at NSF, thus continuing to play significant roles in the chemistry community. As Table I shows, these program officers represented a wide variety of institutions. From 1999 on, each year at least one of the chemistry program officers came from a community college.

Table I. NSF/DUE Chemistry Program Directors

1985-1986	DeWitt Stone (Clemson U)
1986-1988	Nina Roscher (American U, and 1990-1998 in an administrative capacity)
1990-1992	Gene Wubbels (Grinnell College, moved to Washington College, MD)
1990-1992	Curtis Sears (Georgia State U, came back to NSF at intervals through 2011)
1991-1992	John Clevenger (Truckee Meadows Community College)
1992-1994	Stanley Pine (California State U-LA)
1992-2012	Susan Hixson (Mt. Holyoke College, stayed on at NSF as a permanent employee)
1994-1998	Hal Richtol (Rensselaer Polytechnic Institute, stayed on part time 1998-current)
1995-1998	Frank Settle (Virginia Military Institute, moved to Washington and Lee College)
1994-1995	Gene Wubbels (Washington College, moved to U Nebraska-Kearny)
1998-1999	Gary Long (Virginia Tech)
1999-2000	Mel Druelinger (Colorado State U at Pueblo)
1999-2001	Victoria Bragin (Pasadena City College, moved on to win the Van Cliburn International Amateur Piano Competition)
2000-2002	Robert Boggess (Radford U)
2001-2003	Iraj Nejad (Mt. San Antonio College, CA)
2002-2003	Alex Grushow (Rider U)
2003-2004	John Dwyer (College of St. Catherine, MN)
2003-2004	Elizabeth Dorland (Mesa Community College, moved to Washington U)
2004-2006	Kathleen Parson (Macalester College)
2004-2006	Harry Ungar (Cabrillo College)
2006-2008	Pratibha Varma-Nelson (St. Xavier College, moved to Northeastern Illinois U, then moved to Indiana U-Purdue U Indianapolis)
2006-2009	Eileen Lewis (UC Berkeley and Cañada College)
2008-2011	Bert Holmes (U North Carolina-Asheville)

Continued on next page.

Table I. (Continued). NSF/DUE Chemistry Program Directors

2008-2011	Eun-Woo Chang (Truckee Meadows Community College, moved to Montgomery College, MD)
2011-2012	Pamela Brown (New York City College of Technology/CUNY City Tech)
2011-current	Joseph Grabowski (U Pittsburgh)
2012-current	David Brown (Southwestern College)

STEM-Discipline Programs in DUE in 1992

Over the period 1992-2012 the core programs in DUE fell into two categories: those that supported projects aimed at improving student learning in a STEM discipline or interdisciplinary course or curriculum, and those that served a broader goal such as improving the STEM preparation of future K-12 teachers, increasing the number of students graduating with undergraduate STEM degrees, or supporting programs resulting in two-year technology degrees.

In 1992 the programs offered by DUE were solely STEM-discipline ones. Three separate options were available. The Instrumentation and Laboratory Improvement (ILI) program, initiated in 1985, provided support for instruments to be used in laboratories of undergraduate STEM courses. Given the importance of instruments to the field, the chemistry discipline was a major player in the program. Over the next decade the ILI program succeeded in setting new standards for the expected level of instrumentation in first-year general chemistry courses, organic chemistry, and the upper-level courses. For example, the widespread support of nuclear magnetic resonance spectrometers (NMRs) under ILI at a time when NMRs were not yet used in undergraduate laboratories, nor widely perceived as appropriate for such use, undoubtedly laid the groundwork for the later recommendation by the Committee on Professional Training of the American Chemistry Society that all approved undergraduate chemistry programs must have an operational NMR (*5*). The ILI program provided only the matching funds required to purchase an instrument, and the benefit of this restriction was that the program was extremely effective in promoting the spread of instrumentation at a relatively low cost to the NSF. In 1996 alone, a total of 110 chemistry projects were supported for a total of only $3.9 million (*6*). An evaluation (*7*, *8*) showed that the majority of ILI grantees gave presentations at professional meetings or published papers in professional journals, all without financial support from the ILI program. While the ILI projects were awarded on the basis of the educational impact to be expected, most of these impacts were new types of analyses that students were able to perform and resulting increases in the range of investigations that undergraduate students were able to undertake in the laboratory.

The second STEM-discipline program available in 1992 was the Undergraduate Faculty Enhancement (UFE) program, initiated in 1988, that supported workshops or other mechanisms for offering opportunities for faculty to learn about new developments in science or new laboratory techniques. Through 1993, activities of a purely pedagogical nature were not supported under the UFE program. When that restriction was lifted, the first chemistry workshop supported (NSF-DUE-9455107 (*9*)) that had a pedagogical goal was intended to acquaint faculty attendees with the theory and practice of cooperative learning in the general chemistry laboratory. The chapter by Melanie Cooper and Santiago Sandi-Urena (*10*) in this Book provides more information about that project.

The third STEM-discipline program available in 1992 was the Course and Curriculum Development (CCD) program, initiated in 1991, that provided support to revitalize the content, conduct, and quality of undergraduate STEM courses. The program encouraged the design of courses that would enhance interest in STEM, increase the participation of underrepresented students, and encourage the preparation of K-12 teachers. A special concern of the program was with large enrollment courses. Initially most of the chemistry projects focused on the introduction of new science content or the reordering of content within courses or sequences. Others of the early chemistry CCD projects focused on applying newly-available software to applications in chemistry, often developing animations to illustrate chemical concepts and processes.

All three STEM-discipline programs tended to focus on single courses or course sequences at individual institutions.

Introduction of Systemic Approaches in Chemistry

When Dr. Luther S. Williams became Assistant Director for EHR in 1990, he noted that the many smaller reform efforts in the K-12 school systems were introducing chaos into the system, not unity (*11*). He established the Office of Systemic Reform and required that K-12 reform efforts bring together the policy, government, and fiscal components of a K-12 school system.

In line with Dr. Williams' interest in systemic programs, the Systemic Changes in the Undergraduate Chemistry Curriculum (Chemistry Initiative) was launched in 1994 (*12*). The rationales for initiating this program were that chemistry courses are required as part of many STEM majors beyond chemistry and that chemistry courses often prove to be a stumbling point for students in those majors. Thus, significant changes in the chemistry curriculum would directly impact a major portion of undergraduate STEM students. Projects supported through the Chemistry Initiative were allowed to request up to $1 million per year for three to five years in efforts to enhance the learning and appreciation of science through significant changes in chemistry instruction, and to make fundamental changes in the role of chemistry within the institution, including better integration with curricula in related disciplines. Projects were required to involve a coalition of institutions. During 1994-1995, small planning grants were made (*13*), and in 1995-1996, five full projects were supported (*14–16*). Table II lists the five projects, showing for each the lead institution to which each grant was made, the principal investigator, the total award amount for the five-year grant period, the grant number, the title, and a brief description of the goal of the project. Efforts under some of the planning grants that did not receive funding as full projects nevertheless engendered new ideas that went on to secure funding under other programs. A second component of the Chemistry Initiative was an effort in which coalitions of institutions that had not received full award grants could apply for funds to adapt and adopt materials from one or more of the five major projects (*17*, *18*). Since NSF places an emphasis on funding new and innovative work, this Emphasis on Adaption and Adoption opportunity was unique at the agency and recognized the long-standing difficulty in spreading educational efforts beyond their original developers. The expectation was that awarding support directly to potential adaptors would serve as a more effective "pull" mechanism as opposed to the usual practice of giving funds to the developer and asking the developer to "push" dissemination. In 1998-1999 nine awards for a total of $1.4 million were made under the adaption and adoption effort.

Table II. Chemistry Initiative Projects

Institution	Investigator(s)	Amount	Award
Beloit College	Brock Spencer	$2,715,000	NSF-DUE-9455918
ChemLinks Coalition: Making Chemical Connections			
UC-Berkeley	Bradley Moore/ Angelica Stacy	$2,865,000	NSF-DUE-9455924
Sweeping Change in Manageable Units: A Modular Approach for Chemistry Curriculum Reform			
• *To jointly develop complementary innovative modules to use as the basis for a redesigned chemistry curriculum – became ChemConnections*			
U Wisconsin-Madison	John Moore	$3,750,000	NSF-DUE-9455928
Establishing New Traditions: Revitalizing the Curriculum			
• *To design and implement interdisciplinary course clusters; to change fundamentally the ways students, faculty, and administrators view their roles – generated Process Oriented Guided Inquiry (POGIL)*			
CUNY City College	David Gosser	$1,525,000	NSF-DUE-9455920
A Workshop Chemistry Curriculum			
• *To modify course structures to include student-led workshops and mentorship by recent course graduates – became Peer-Led Team Learning (PLTL)*			
UC-Los Angeles	Orville Chapman	$2,150,000	NSF-DUE-9555605
Molecular Science			
• *To design a curriculum to cut across departmental and disciplinary lines to include all activities that involve the study of atoms and molecules, with a particular emphasis on environmental science, materials science, and molecular life science – generated Calibrated Peer Review (CPR)*			

Looking back a decade after the conclusion of the Chemistry Initiative, a number of conclusions can be drawn. First, even the proposed projects did not intend to focus on the entire undergraduate chemistry curriculum, and they did not expect to make significant changes in the relationship of the chemistry curriculum with the curricula in the other STEM disciplines. The limit of $1 million a year for requests did not allow faculty to propose projects of this scope. Instead, most of the proposed projects focused on making significant changes in the first or first and second years of the chemistry curriculum. Second, many of the materials and pedagogies developed under the grants were not fully perfected and disseminated within the lifetime of the awards. Five years turned out to be too short a period to develop materials and pedagogies and then run several cycles evaluating the implementations and revising the materials and pedagogies in light of the findings from the evaluations. Third, although a reasonable fraction of the materials and pedagogies that were developed did not persist, and others did not see widespread use, some elements of the materials and pedagogies have become well-established across the nation within the undergraduate chemistry curriculum, and most of these also have moved into use in other disciplines. For example, the ChemConnections modules are in continued use in a limited number of institutions; Process Oriented Guided Inquiry (POGIL) developed under the New Traditions project continues to expand its use in chemistry courses and in other disciplines (*19*), Peer-Led Team Learning (PLTL) developed under the Workshop Chemistry project is used widely across hundreds of institutions (see the chapter by Pratibha Varma-Nelson (*20*) in this Book), and Calibrated Peer Review (CPR) developed under the Molecular Science project is well-established (see the chapter by Arlene Russell (*21*) in this Book). All of these have received additional funding from other sources since the formal end of the Chemistry Initiative. However, the high level of concentrated funding under the Initiative provided a base for developing each effort, provided for several rounds of evaluation and modification, and, importantly, brought together national networks of faculty that were using the materials and pedagogies and were invested in their further development (*22*).

Along different lines, the Chemistry Initiative proved to be significant for the development of the undergraduate chemistry education community. The five major projects included faculty from more than 70 institutions working for five years with access to more than $14 million. Significant communication occurred within the five projects, and a later workshop program offered jointly by the five projects led to on-going interactions and collaborations across the five projects and provided access to the materials and pedagogies for hundreds of additional faculty. Important results for the chemistry education community included the formation of new collaborations among faculty across a wide variety of institutions; the ability of faculty to investigate research questions and to develop and implement active learning strategies to a degree not possible earlier under smaller more isolated projects; the development of assessment instruments, again not possible under smaller, isolated projects; and the training of faculty, graduate students, and postdoctoral fellows in education research methodology, pedagogy, and evaluation, again because the size and scope of the projects provided suitable environments and opportunities.

Prior to the work of the Chemistry Initiative, the chemistry community for decades had included an active education research community. However, this community was relatively small, and the crossover of the work done by this research community with other curriculum and teaching efforts in the broader chemistry community was not routine. The Chemistry Initiative brought together large networks of faculty, almost none of whom initially self-identified as chemistry education researchers. As the projects proceeded and the materials and pedagogies ran into challenges, faculty began to realize the value of education research findings that could assist them in making revisions to their materials. Since the NSF had not required that a research expert be a part of the project team, often the project evaluators were the ones who played the role of steering the faculty to the research literature. Over the lifetime of the Initiative, hundreds of chemistry faculty became aware of the research literature and its value. Some faculty moved into the chemistry education research field or chose to become trained in evaluation methods. For an example of a faculty member who participated in a Chemistry Initiative project and then moved into a career in chemistry education research, see the chapter in this Book by Stacey Lowery Bretz (*23*). Many other faculty realized the value of consulting the research literature before they set forth to develop classroom interventions. Thus, the Chemistry Initiative went a long way towards changing the culture in chemistry education from a situation where faculty would write new content materials or develop new software based only on their own content knowledge to one where faculty would consult the literature or colleagues to see what learning issues should be considered along with the content issues.

The Chemistry Initiative was unique to the chemistry discipline and was not repeated in other disciplines. An earlier Engineering Education Coalition (EEC) program had been run out of the Directorate for Engineering at NSF. In retrospect the impacts of the EEC on the engineering discipline in terms of catalyzing the development of faculty networks and greatly enhancing the role of engineering education research can be seen to be similar to those of the Chemistry Initiative on the chemistry discipline.

STEM-Discipline Programs in DUE: Later Developments

While the Chemistry Initiative was in progress, the regular STEM-discipline programs continued for all of the STEM disciplines, and chemistry faculty continued to participate in the ILI, UFE, and CCD programs. See the chapter by Thomas Holme and Heather Caruthers (*24*) in this Book for a description of the evolution of a CCD project initially funded during this period.

In 1999, the existing three STEM-discipline programs were combined to form the Course, Curriculum and Laboratory Improvement (CCLI) program. The CCLI program (NSF 00-63) included tracks for materials development, large-scale national dissemination (workshops) efforts, and an adaptation and implementation (A&I) track that was designed to promote the dissemination of projects that had been effective at other institutions. All tracks allowed requests for salaries, travel, instruments, workshops for dissemination, supplies, and indirect costs. At the same time, expectations increased that the proposed projects would be grounded in the knowledge base and would include a meaningful evaluation. Although these expectations existed in the earlier programs, they had been less emphasized, particularly in the ILI program that provided no financial support for evaluation. Most of the A&I proposals continued to ask for instrumentation and generally asked only for funding for the instrument. With the new flexibility in the program, CCLI proved to be an ideal source of funding for many of the extended efforts that grew out of the Chemistry Initiative.

In 2001 an explicit program, Assessment of Student Achievement (ASA, NSF 01-82), was offered as a response, first, to the increasing emphasis by the regional higher education accreditation agencies on institutional assessment of learning and, second, the need for better assessment tools that would reflect the goals and emphases in the new courses and curricula that were being developed across the STEM fields. In the early years the response by the STEM community was strongest for development of assessment methods for courses and curricula. In the first round of the ASA competition the response was particularly strong from the chemistry community, and three projects were supported: *ChemQuery: An Assessment System for Mapping Student Understanding in Chemistry* (NSF-DUE-0125651); *Technology Based Chemistry Assessments* (NSF-DUE-0126050); and *Real-Time Multi-Dimensional Assessment in General Chemistry* (NSF-DUE-0127650). Over the next decade the ASA program became a track within the CCLI program, and eventually assessment became one emphasis among others encouraged in the CCLI program. Throughout this decade assessment efforts expanded within the chemistry discipline, and the American Chemical Society Exams Institute under the leadership of Thomas Holme was one of the key players in such assessment efforts with a number of projects including NSF-DUE-0717769, NSF-DUE-0817409, and NSF-DUE-0920266. For further examples of assessment efforts in chemistry, see the chapter by Pam Paek and Thomas Holme (*25*) in this Book, the chapter by Kristen Murphy and April Zenisky (*26*) in this Book, and the chapter by Stacey Lowery Bretz (*23*) in this Book.

Meanwhile, in 2005 the CCLI program evolved from including separate tracks to a program designed around a set of components that could be combined to develop projects of various sizes and scopes (NSF 05-559). The intent was to stress the various steps that are required to develop, implement, evaluate, and disseminate an educational innovation, but no one project was expected to carry out all of the steps. The components included conducting research on undergraduate STEM teaching and learning, creating learning materials and teaching strategies, developing faculty expertise, implementing educational innovations, and assessing learning and evaluating innovations. The program again increased the emphasis on building on prior work in the field and contributing to the knowledge base of undergraduate STEM education research and practice, expected projects to contribute to building a community of scholars, and required projects to explicitly identify a set of measurable outcomes to be used in the project management and evaluation. As proposers became acquainted with these requirements, and as reviewers took these requirements more seriously, successful proposals became more sophisticated in terms of teaching and learning issues and evaluation, and the program became less accessible to those without a background in education work. For example, in the early 1990's a project could be funded that proposed only the development of an animation, but by 2005-2010 simulations were expected to have demonstrable impacts on student learning. Many fewer instruments were requested and funded. (The numbers of proposals submitted by chemistry departments to ILI and CCLI, the types of instruments requested, and the success rates of the declining numbers of instrument proposals have been compared for the periods 1996-1998 versus 2006-2008 (*27*).) On the other hand, proportionately more work in discipline-based education research was funded, such as that detailed in the chapter by Dawn Rickey and Lydia Tien (*28*) in this Book.

Finally, CCLI was renamed Transforming Undergraduate Education in Science, Technology, Engineering and Mathematics (TUES) in 2010 in accordance with the NSF Strategic Plan for FY 2006-2011 (NSF 06-48) that emphasized that NSF advances scientific discovery by supporting transformational capabilities. The name change also was intended to highlight to the community that CCLI/TUES projects should be considered as elements of a large, nationwide effort to transform the undergraduate STEM experience.

Summary of Chemistry in the STEM-Discipline Programs

The chemistry community has been extremely active in using the opportunities available under ILI, CCD, UFE, the Chemistry Initiative, CCLI, and TUES. At the same time, the changing program emphases moved the chemistry community into new areas and expertises over the period 1992-2012. From a focus in 1992 on instruments and chemistry content, by 2012 faculty involved in undergraduate chemistry education were focusing on content, pedagogy, cyber applications, assessment, chemistry education research, and evaluation. By 2012 faculty were working across more institutions and in more collaborations, and were undertaking more complex projects in which content, pedagogy, and education research all played roles. In addition, as described below, chemists also were working on projects that had broader goals than those emphasized under ILI, CCD, UFE, CCLI, and TUES.

Broader-Based Programs in DUE Relevant to Chemistry

The first program in DUE to respond to goals beyond the improvement in student learning in STEM disciplines was the Collaboratives for Excellence in Teacher Preparation (CETP) program begun in 1993. This program sought to reform PreK-12 teacher preparation through meaningful collaborations between STEM disciplinary and education departments at universities and colleges. The goal of CETP was to improve the science, mathematics and technology preparation of future K12 teachers through efforts by coalitions of higher education institutions and K-12 schools and school districts and other stakeholder organizations. This systemic focus was in line with the EHR Directorate's interest in the early 1990's in undertaking systemic efforts. In the early 2000's the CETP program was phased out, and the Robert Noyce Teacher Scholarship (Noyce) program was begun that provided funds to institutions of higher education to support scholarships, stipends, and teacher preparation programs for STEM majors who committed to teaching in high need K-12 schools (NSF 13-526). CETP and Noyce brought additional chemistry faculty, beyond those working in the discipline-specific DUE programs, to reform efforts, encouraged chemistry faculty to work on issues beyond the chemistry discipline, and catalyzed changes in the chemistry curriculum that brought best practices to the undergraduate classrooms of future K-12 teachers. For example, Joseph Heppert aligned his work under CETP award NSF-DUE-9876676 and Noyce grant NSF-DUE-0934906 with his efforts to develop a program that allows students in Kansas to earn both a degree in STEM and a teaching license in four years. As described in a chapter in this Book, Donald Wink, Sharon Fetzer Gislason, and Julie Ellefson (*29*) used work begun under a CETP grant to establish a set of courses for pre-elementary education majors.

In 1994 the Advanced Technological Education (ATE) program was initiated, and this program has remained a large and well-funded program (NSF 11-692). The goal of ATE was to promote improvement in advanced technological education at the national and regional levels through support of curriculum development and program improvement at the undergraduate (community college) and secondary school levels, especially for technicians being educated for the high performance workplace of advanced technologies. The American Association of Community Colleges was the force behind securing funds from Congress to initiate the program, and ATE remains the only program at NSF targeted specifically to community colleges. For the roughly half of all community colleges that engage in explicit technician training, ATE has been a major catalyst for the updating of their technician programs, and national ATE Centers have provided concentrated resources available to individual community colleges and their faculty in specific fields of STEM advanced technology. For example, the National Network for Pulp and Paper Technology Training (NSF-DUE-0902811) and the Northeast Biomanufacturing Center and Collaborative (NSF-DUE-0903208) each are partnerships among community colleges, universities, secondary schools, and industries to enhance technician training important for chemical industries. Many of the smaller ATE projects resemble those funded under CCLI/TUES in that course and curricular changes focus on materials development, pedagogical changes, and evaluation. Although not all community colleges engage in technician training, ATE nevertheless has served as a mobilizing force for engaging all such institutions in NSF programs through outreach to faculty, training sessions for grants officers and community college administrators, and other efforts. In addition, ATE has co-funded projects such as those in the Chemistry Initiative in order to bring increased involvement of community college faculty to these efforts.

The National STEM Education Digital Library/Distributed Learning (NSDL) program was initiated in 2000 with the goal of establishing a virtual library that would enable the discovery, creation, collection, organization, and delivery of quality STEM teaching and learning resources appropriate for educators and learners at all levels (NSF 10-545). As a part of this effort, NSDL supported the development of online collections of materials in specific STEM areas. One of the highly successful chemistry collections is the Analytical Sciences Digital Library (ASDL) that includes a collection of electronic resources for teachers, students, and practitioners interested in chemical measurements and instrumentation and that builds a community of users through various communication efforts. In the late 1990's program officers at the NSF in DUE and the Division of Chemistry jointly held two workshops to encourage faculty and representatives from industry to consider the need for reforms in the undergraduate analytical sciences curriculum. These workshops culminated in a report, *Curricular Developments in the Analytical Sciences* (*30*), that included recommendations for future actions. As a result, a number of symposia were begun at professional meetings to encourage and highlight curricular reforms in analytical chemistry, and in 2001 with the advent of the NSDL program Theodore Kuwana and Cynthia Larive undertook the development of the ASDL (NSF-DUE-0121518 and (*31*)). A second project of importance to the chemistry community is the ChemEd Digital Library, a joint project of the *Journal of Chemical Education* and the American Chemical Society (NSF-DUE-0632303, NSF-DUE-0632247, and NSF-DUE-0632303 and (*32*)) that collects digital resources, tools, and online services that enhance the teaching and learning of chemical science and makes them publicly available. Thus, NSDL has played an important role in supporting the generation, collection, and dissemination of important materials to the chemistry community.

In 2002, under the leadership of Division Director Norman Fortenberry, DUE and two research directorates ran a single joint competition (NSF 02-038) of the Higher Education Centers for Learning and Teaching. Each multi-institutional project was to develop a center that would provide a nucleus for coordinated efforts to reform teaching and learning at the nation's colleges and universities through a blend of research, faculty professional development, and education practice. Two $10-million, five-year Centers were funded, one joint with the Directorate for Engineering and the second joint with the Directorate for Mathematical and Physical Sciences. This second Center, the Center for the Integration of Research, Teaching and Learning (CIRTL), has worked for ten years to develop, implement, and spread an innovative model for strengthening the traditional graduate training at major research universities so that the STEM students receive significant experience with educational issues. The long-term results are expected to include beginning faculty at all types of institutions, including new chemists, who start their faculty careers with an increased ability to develop effective learning environments in their undergraduate classes. The chapter by Robert Mathieu (*33*) in this Book provides more information about the CIRTL project.

In 2002, the Science, Technology, Engineering, and Mathematics Talent Expansion Program (STEP) was initiated with the goal of increasing the number of students (U.S. citizens or permanent residents) pursuing and receiving associate or baccalaureate degrees in established or emerging fields within STEM (NSF 11-550). The Louis Stokes Alliances for Minority Participation (LSAMP) program, managed by the Division of Human Resource Development within EHR, served as a conceptual precursor for STEP. Initiated in 1990, LSAMP strengthened and encouraged STEM baccalaureate degree production of students from underrepresented populations (NSF 11-543), while under STEP this focus was extended to all students and to community college degrees. The funded projects in STEP generally work across all of the engineering fields or across all of the physical and natural sciences. The field of chemistry has been involved in important ways. A number of projects have chemists as principal or co-principal investigators, and these chemistry faculty thus broaden their experience and influence at the institution to disciplines beyond chemistry and to issues of student recruitment and retention. See the chapter by Maureen Scharberg (*34*) in this Book for an example of such an award. In other cases, chemistry materials such as those developed under PLTL are being spread to other STEM disciplines in order to take advantage of the known ability of PLTL to increase retention within courses. Finally, findings from STEP on effective mechanisms for the recruitment and retention of undergraduate students are likely to be of significance to chemistry departments concerned with these issues. For example, many STEP projects are investigating the use of early undergraduate research to recruit and retain students in STEM or to engage students early on while they still may be struggling with strengthening their STEM backgrounds. The results from these efforts should help to inform the practice of undergraduate research within chemistry departments.

In 2006 chemistry became an allowed discipline in the NSF Scholarships in Science, Technology, Engineering, and Mathematics (S-STEM) program (NSF 12-529) that replaced the NSF Computer Science, Engineering, and Mathematics Scholarships (CSEMS) program that did not include participation by chemistry. The S-STEM program makes grants to institutions of higher education to support scholarships for academically talented, financially needy students, enabling them to enter the workforce following completion of an associate, baccalaureate, or graduate level degree in science and engineering disciplines. The awards allow relatively little funding for purposes beyond the scholarships, so S-STEM awards are not important sources of funding for curricular developments or other programmatic developments. However, providing direct financial support to students often enhances other efforts, such at STEP projects or undertakings that may be more successful when students are able to move from time spent on earning money to time freed for academic pursuits.

Trajectory of Changes in DUE 1992-2012

This article summarizes the evolution of programs in DUE over the period 1992-2012. Some of these changes were conceived and put in place by the Division Directors and/or the program officers in DUE. Others were mandated by Congress (ATE, Noyce, STEP, S-STEM) with DUE developing only some aspects of the programs. A continuing concern was to increase the participation of community colleges and their faculty in DUE programs. The chapter by Thomas Higgins (*4*) in this Book outlines the importance of the community college community to the STEM undergraduate enterprise. Although relatively few proposals have come from such institutions to the STEM-discipline programs in chemistry or in the other STEM disciplines, some successful efforts have been supported at community colleges. The chapter in this Book by Donald Wink, Sharon Fetzer Gislason, and Julie Ellefson (*29*) provides an example of such efforts. On the other hand, community colleges have shown strong participation in ATE, STEP, and S-STEM.

All of the projects that have been funded in all of the programs offered by DUE were proposed and carried out by members of the STEM community, and it is their work that provides the changes that have been catalyzed by the programs in DUE. The other chapters in this Book provide details of how specific projects or groups of projects have succeeded in generating reforms in the undergraduate chemistry environment.

References

1. The period 1992-2012 also represents the years that the author served as a program officer in DUE at the NSF.
2. Walsh, J. *Science (Washington, DC)* **1986**, *231*, 911–912.
3. Citations of the form NSF XX-XX or NSF XX-XXX refer to program solicitations or reports that can be found on the NSF Web site at http://www.nsf.gov. For program solicitations, earlier versions of the solicitation can be located by noting the "Replaces Document(s)" notation that appears under the solicitation number on the first page of the solicitation.
4. Higgins, T. In *Trajectories of Chemistry Education Innovation and Reform*; Holme, T., Cooper, M. M., Varma-Nelson, P., Eds.; ACS Symposium Series 1145; American Chemical Society: Washington, DC, 2013; Chapter 12.
5. American Chemical Society Committee on Professional Training, *Undergraduate Professional Education in Chemistry: ACS Guidelines and Evaluation Procedures for Bachelor's Degree Programs*; American Chemical Society: Washington, DC, 2008.
6. J. Chem. Educ. Report: NSF instrumentation and laboratory improvement grants in chemistry. *J. Chem. Educ.* **1997**, *74*, 29–32.
7. Tuss, P.; Frechtling, J.; Ewing, T. *A Report on an Evaluation of the National Science Foundation's Instrumentation and Laboratory Improvement Program*, Westat, Inc.: Rockville, MD, 1998. http://www.nsf.gov/pubs/1998/nsf9833/ili-toc.htm (accessed February 2013).
8. Moore, J. W. *J. Chem. Educ.* **1998**, *75*, 799.
9. Citations of the form NSF-DUE-YYYYYYY refer to grants awarded by NSF. The abstract, contact information for the Principal Investigator, and other information about the award can be found by using the Search Awards feature on the NSF Web site at http://www.nsf.gov. Enter only the numerical portion of the award citation, and check active or expired awards as appropriate when making the search.
10. Cooper, M. M.; Sandi-Urena, S. In *Trajectories of Chemistry Education Innovation and Reform*; Holme, T., Cooper, M. M., Varma-Nelson, P., Eds.; ACS Symposium Series 1145; American Chemical Society: Washington, DC, 2013; Chapter 4.

11. Novak, C. *Interview with Luther S. Williams*; TECHNOS Quarterly, **1995**, *4*. http://www.ait.net/technos/tq_04/4williams.php (accessed February 2013).
12. Ross, L. *Chem. Eng. News* **1993**, *71*, 4–5.
13. Ross, L. *Chem. Eng. News* **1994**, *72*, 25.
14. Raber, L. *Chem. Eng. News* **1995**, *73*, 7.
15. Hixson, S. H.; Sears, C. T., Jr. *J. Chem. Educ.* **1995**, *72*, 639–649.
16. Hixson, S. H.; Sears, C. T., Jr. *J. Chem. Educ.* **1996**, *73*, 525.
17. Chem Eng. News. *Chem. Eng. News* **1996**, *74*, 17.
18. Hixson, S. H.; Richtol, H. H.; Settle, F. A. *Chem. Educ.* **1997**, *1*, 1–3.
19. POGIL. http:/www.pogil.org/ (accessed September 2013).
20. Varma-Nelson, P.; Banks, J. In *Trajectories of Chemistry Education Innovation and Reform*; Holme, T., Cooper, M. M., Varma-Nelson, P., Eds.; ACS Symposium Series 1145; American Chemical Society: Washington, DC, 2013; Chapter 7.
21. Russell, A. In *Trajectories of Chemistry Education Innovation and Reform*; Holme, T., Cooper, M. M., Varma-Nelson, P., Eds.; ACS Symposium Series 1145; American Chemical Society: Washington, DC, 2013; Chapter 9.
22. Yarnell, A. *Chem. Eng. News* **2002**, *80*, 35–36.
23. Bretz, S. L. In *Trajectories of Chemistry Education Innovation and Reform*; Holme, T., Cooper, M. M., Varma-Nelson, P., Eds.; ACS Symposium Series 1145; American Chemical Society: Washington, DC, 2013; Chapter 10.
24. Holme, T.; Caruthers, H. In *Trajectories of Chemistry Education Innovation and Reform*; Holme, T., Cooper, M. M., Varma-Nelson, P., Eds.; ACS Symposium Series 1145; American Chemical Society: Washington, DC, 2013; Chapter 5.
25. Paek, P; Holme, T. In *Trajectories of Chemistry Education Innovation and Reform*; Holme, T., Cooper, M. M., Varma-Nelson, P., Eds.; ACS Symposium Series 1145; American Chemical Society: Washington, DC, 2013; Chapter 11.
26. Zenisky, A.; Murphy, K. In *Trajectories of Chemistry Education Innovation and Reform*; Holme, T., Cooper, M. M., Varma-Nelson, P., Eds.; ACS Symposium Series 1145; American Chemical Society: Washington, DC, 2013; Chapter 6.
27. Hixson, S.; Chang, E.-W.; Holmes, B. E. *J. Chem. Educ.* **2010**, *87*, 247–249.
28. Rickey, D.; Tien, L. In *Trajectories of Chemistry Education Innovation and Reform*; Holme, T., Cooper, M. M., Varma-Nelson, P., Eds.; ACS Symposium Series 1145; American Chemical Society: Washington, DC, 2013; Chapter 3.
29. Wink, D.; Fetzer Gislason, S.; Ellefson, J. In *Trajectories of Chemistry Education Innovation and Reform*; Holme, T., Cooper, M., Varma-Nelson, P., Eds.; ACS Symposium Series 1145; American Chemical Society: Washington, DC, 2013; Chapter 8.
30. *Curricular Developments in the Analytical Sciences, A Report from the Workshops, 1996 and 1997.* http://www.asdlib.org/files/curricularDevelopment_report.pdf (accessed February 2013).
31. Analytical Sciences Digital Library. http://www.asdlib.org/ (accessed February 2013).

32. Chemical Education Digital Library. http://www.chemeddl.org/ (accessed February 2013).
33. Mathieu, R. In *Trajectories of Chemistry Education Innovation and Reform*; Holme, T., Cooper, M. M., Varma-Nelson, P., Eds.; ACS Symposium Series 1145; American Chemical Society: Washington, DC, 2013; Chapter 13.
34. Scharberg, M. In *Trajectories of Chemistry Education Innovation and Reform*; Holme, T., Cooper, M. M., Varma-Nelson, P., Eds.; ACS Symposium Series 1145; American Chemical Society: Washington, DC, 2013; Chapter 14.

Trajectories of Reform in Curricula

Chapter 3

Research on Learning in the Chemistry Laboratory

A Trajectory Connecting Student Outcomes to Thinking Processes

Dawn Rickey*[,1] and Lydia T. Tien[2]

[1]Department of Chemistry, Colorado State University, Fort Collins, Colorado 80523-1872
[2]Department of Chemistry and Geosciences, Monroe Community College, 1000 E Henrietta Rd, Rochester, New York 14623
***E-mail: dawn.rickey@colostate.edu**

This chapter describes how our research on student learning in the undergraduate chemistry laboratory unfolded over the past two decades, including example results. We discuss the evolution of our work examining student learning in the context of an instructional method we developed called the Model-Observe-Reflect-Explain (MORE) Thinking Frame. Our research trajectory ultimately connects aggregate student learning outcomes to the thinking processes that individual students engage in while participating in guided discovery.

This chapter describes how our research on student learning in the chemistry laboratory has evolved over the past two decades, including some example results. Our journey began in graduate school at the University of California at Berkeley during the early 1990s, during the time when the five systemic change initiatives in undergraduate chemistry education (*ModularChem Consortium, ChemLinks, Molecular Science, New Traditions, and Workshop Chemistry*) were funded by the National Science Foundation. While our collective experiences include work with four out of five of these initiatives, the bulk of our efforts has focused on research and development in the context of the guided-discovery pedagogy known

© 2013 American Chemical Society

as the Model-Observe-Reflect-Explain (MORE) Thinking Frame (*1–3*). Thus, here we discuss the trajectory of our research on student learning in the context of MORE laboratory instruction, which ultimately connects aggregate student learning outcomes to the thinking processes that individual students engage in while participating in guided discovery.

We begin by outlining our design principles and the implementation of the original MORE curriculum and instruction. Next, we describe some of the findings from our early research studies that compared the learning outcomes of a MORE treatment group with a control group. Finally, we connect the early work to our more recent research endeavors focused on the relationships between cognition of individual students and the thinking processes that are associated with students' success at applying the scientific models they develop effectively in new contexts.

Our original instructional design principles for Model-Observe-Reflect-Explain instruction were garnered from previous research in chemistry education as well as the broader science education literature (e.g., (*4–18*)) and included: promote metacognition (*19*); support guided discovery; and engage students in authentic scientific inquiry. These principles guided the development of a modular, first-semester general chemistry laboratory curriculum and instruction. The cognitive aspects of these design elements were integrated through the use of the MORE Thinking Frame, as well as other curricular and instructional supports.

In the original MORE curriculum, each laboratory module began with a general question to be answered over three or four weeks. The curriculum focused on investigating the chemistry of three biologically-relevant systems: *Detecting and discriminating odors: How does the nose know?; Designing an effective antacid: How do YOU spell relief? (2)*; and *Investigating sunscreen effectiveness: How can you avoid getting burned? (1)* These modules focused on models for structure and bonding, acid-base equilibria, and the interaction of light with matter, respectively.

Students carried out several experiments within each module; and each experiment corresponded to one iteration of the MORE Thinking Frame. At the beginning of each module, students were asked to write their initial *model* of how they thought the chemical system under study would function. Each time they conducted an experiment (*observe*), the students were prompted to *reflect* on the experiment (during and following the experiment), and to *explain* their results. In each successive week, students refined their models based on the observations, reflections, and explanations of the previous week(s), and then progressed to another experiment related to the overarching laboratory module question. The MORE Thinking Frame is a representation of the thought processes of scientists that encourages students to reflect upon their own understanding of chemistry throughout the laboratory program and to revise their ideas based upon the empirical evidence they collect.

Students participating in MORE were given specific experimental questions to study in the initial weeks of each laboratory module, but by the end of each module increased responsibility was placed on the students. As the students proceeded through the semester, they progressed from designing a simple experiment given a set of experimental questions from which to choose in the first module to formulating their own experimental question, protocol, and approach to data analysis in the third module. At the end of each module, students presented their results in oral and/or written formats and critiqued their peers' methods and data analyses. Further details regarding the original implementation of the MORE Thinking Frame can be found in references (*1–3*).

Comparing the Learning Outcomes of MORE Students and Control Students

Our original research design compared several learning outcomes for groups of students participating in MORE versus standard laboratory curricula and instruction. Using a within-course experimental variation, the MORE curriculum and instruction was implemented in two laboratory sections selected at random from the forty-six sections in the general chemistry course of a large research university. An experienced graduate student instructor who was involved in the development of MORE taught the MORE laboratory sections. Two other experienced graduate student instructors who were enthusiastic about teaching laboratory sections in the standard way were chosen to teach Control sections. Each Control group was selected randomly from sections meeting at the same time as each of the MORE sections. Students in all laboratory sections attended lectures given by the same professor. There were no statistically-significant differences in the means of the MORE and Control groups on various pre-course measures. These included Math SAT scores, number of previous chemistry courses taken, scores on a test of chemistry concepts given during the first week of classes, and gender distribution.

The Control laboratory curriculum was comprised of eight one– or two–week laboratory experiments and did not employ the MORE Thinking Frame. Like the MORE curriculum, the majority of the laboratory experiments for the Control sections employed a meaningful context such as water chemistry and the thermochemistry of foods. In addition, the Control students were given some limited experimental design opportunities. In contrast to MORE instruction, these opportunities did not allow the Control students to formulate new experimental questions or to build upon experience and understanding gained from investigations they had carried out in previous weeks. Finally, rather than emphasizing an evolving conceptual understanding of a system, the Control pre- and post-laboratory assignments focused primarily on procedures, calculations, and pre-defined data analyses.

We used a combination of qualitative and quantitative methods to characterize and compare student learning outcomes. The data we collected included pre- and post-surveys of attitudes, beliefs, perceptions, and demographics; videotapes of student groups working in the laboratory; lecture-based course examinations; other written assessments; and pre- and post-interviews exploring inquiry, problem solving, and understanding of acid-base chemistry. Here, we present some examples of aggregate student learning outcomes based on analyses of the surveys, regular course examinations, other written assessments, and interviews. Further details regarding these analyses and results can be found in references (*1*) and (*2*).

Example Learning Outcomes Based on Analyses of Survey Data: Student Perceptions of Learning and Scientific Research

Analyses of survey data revealed differences between the MORE (N = 39) and Control (N = 38) students in their perceptions of learning in the laboratory and their beliefs about scientific research at the end of the semester.

For example, in one set of post-survey questions, students were given the list of learning goals shown in Table I. They were asked to respond on a Likert scale from 1 (least important) to 5 (most important) to the question "How do you think [course lecture professor] would rank the importance of each of the following goals for learning in the laboratory?" Analyses of students' responses revealed that the only statistically-significant difference between the MORE and Control groups (two-tailed, unpaired t-test $p < 0.05$) was for the item "Students should learn to relate chemistry to real-world problems." The Control group perceived that the course professor would consider relating chemistry to real-world problems significantly more important than the MORE group did. For each of the other eight learning goals, however, students in the MORE and Control groups agreed on their importance to the course professor. The learning goals are listed in Table I in the order that the entire group (on average) perceived that the course professor would rank them, with the most important goal, to understand chemistry concepts, at the top of the table. It is interesting to note that the students perceived activities associated with a traditional chemistry laboratory course, such as learning to perform laboratory techniques and apply mathematical formulas, as the least important to the course professor.

Next, using the Likert scale, students were asked to rank the same learning goals in terms of what they thought they had actually learned in the laboratory (Table II). The MORE students rated five of the nine items, numbered 1, 2, 3, 4, and 7 in Table II, significantly higher than the Control group (two-tailed, unpaired t-test $p < 0.05$). These items included understanding and presenting understanding of chemistry ideas, and inquiry-related items such as thinking about the meaning of data and learning how chemistry research is done. In contrast, the Control students felt most strongly that they had learned to perform laboratory techniques, the only item receiving a mean score greater than 4. This is consistent with a review of research on laboratory learning (*20*) that concluded that the main thing students learn from standard laboratory courses is laboratory technique. The only item that Control students rated significantly higher than MORE students was the application of mathematical formulas, another activity associated with traditional laboratory courses.

Despite agreement on what the lecture professor valued, the MORE and Control groups differed significantly in what they thought they had learned in their laboratory environments. The items that the Control students rated the highest were activities associated with traditional laboratory courses; these learning goals did not coincide with those that students believed were most important to the professor. In contrast, the MORE students believed that their laboratory environment fostered understanding of chemistry ideas and supported the process of scientific inquiry.

Table I. Summary of responses to the post-survey question "How do you think [course lecture professor] would rank the importance of each of the following goals for learning in the laboratory?"

Students should learn...	*Control Mean*	*MORE Mean*	*Control Rank*	*MORE Rank*
1...to understand chemistry concepts better through hands-on experience.	4.66	4.38	1	2
2...to think about the meaning of data.	4.37	4.46	2	1
3...to relate chemistry to real-world problems.	4.35	3.85	3	3
4...to present my understanding of chemistry to others (written or oral).	3.63	3.66	4	4
5...to find an answer to an experimental question.	3.51	3.36	5	7
6...to work together in groups.	3.38	3.47	6	6
7...how chemistry research is done.	3.21	3.50	8	5
8...to perform laboratory techniques.	3.29	3.33	7	8
9...to apply mathematical relationships (formulas) to an experimental system.	2.91	2.74	9	9

A second set of survey questions probed students' ideas about the activities involved in scientific research. One question, administered on both the pre- and post-survey, presented students with a list of four research activities: formulating an experimental question, following a set of procedures, analyzing data, and drawing conclusions. Students were asked "What activities, if any, do you think are missing from this list?" Analyses of student responses to the pre-survey question showed no significant differences between the MORE and Control groups. The most common responses of both student groups were making a hypothesis about the experiment's outcome (27%); performing experiments, gathering and recording data (25%); and developing experimental procedures (24%).

On the post-survey, the Control students' responses to the question were virtually identical to their answers on the pre-survey. However, the responses of the MORE group were significantly different on the post-survey. On the post-survey, several new responses not present in the pre-survey answers emerged. These included explicating one's initial understanding; reflecting, sense making, or thinking about connections; explaining results to others; and critiquing or thinking about how to improve an experiment. Note that the first three of these new categories correspond to the model, reflect, and explain aspects of the

MORE Thinking Frame, respectively. On the post-survey, significantly greater proportions of MORE students (Fisher's exact test $p < 0.05$) included each of these three aspects compared with the responses of the Control students. In addition, the most common response of the MORE students on the post-survey question (39%) was that refining one's understanding should be mentioned in the list of research activities. This suggests a recognition of the central purpose of scientific research not present in the Control group. Although it is not surprising that the students who were taught to use the Model-Observe-Reflect-Explain Thinking Frame mentioned these aspects more frequently compared with students who did not use this instructional tool, it is important to note that the students reported them in this context as *research activities*. This suggests that the students considered the thinking processes embodied in the MORE Thinking Frame not only as instructional activities for their course, but also as an integral part of scientific research.

Table II. Summary of responses to the post-survey question "How would you rank each of the following in terms of what you learned in the laboratory?"

I learned...	*Control Mean*	*MORE Mean*	*Control Rank*	*MORE Rank*
1...to understand chemistry concepts better through hands-on experience.	3.67	4.26	5/6	3
2...to think about the meaning of data.	3.92	4.54	2	1
3...to relate chemistry to real-world problems.	3.77	4.33	4	2
4...to present my understanding of chemistry to others (written or oral).	2.72	4.24	9	4
5...to find an answer to an experimental question.	3.67	3.95	5/6	5
6...to work together in groups.	3.64	3.92	7	6/7
7...how chemistry research is done.	2.97	3.59	8	8
8...to perform laboratory techniques.	4.28	3.92	1	6/7
9...to apply mathematical relationships (formulas) to an experimental system.	3.78	2.95	3	9

Overall, student responses to the survey questions attest that the MORE students perceived that the elements of the intended instructional design were implemented in their laboratory course. The students indicated that their course supported them in engaging in inquiry. In addition, the MORE students were aware of the importance of refining understanding in scientific research.

Example Learning Outcomes Based on Assessments of Inquiry Skills

We also investigated the development of MORE and Control students' inquiry skills via written experimental design assessments and interviews. Example outcomes are presented for each below.

Written Assessment of Experimental Design Skills

Students completed written pre- and post-experimental design assessments. The pre-assessment asked students to design an experiment to determine if a certain cereal brand contained twice as much iron as a competing brand. The post-assessment asked students to design an experiment to determine if all, some, or none of a spoonful of sugar dissolved in a saturated salt solution. For each assessment, students earned scores in two categories: (1) *protocol development*, which refers to the extent that implementation of the proposed experimental procedure would provide data to answer the posed question; and (2) *data analysis and explanation*, which reflected the quality of a student's description of the planned analysis and clarification of its potential implications. A student's total score on each assessment was the sum of the scores for each category. For the students who completed both the pre- and post-assessments [N(MORE) = 36, N(Control) = 22], analyses of covariance (ANCOVA) using the total score on the pre-experimental design task as the covariate revealed a significant difference in performance between the MORE and Control groups on the post-experimental design task total score ($p < 0.05$). The MORE students significantly outperformed the Control students in the data analysis and explanation subcategory ($p < 0.05$), but not in the protocol development subcategory. In addition, previous chemistry background did not significantly influence experimental design scores.

Interview-Based Assessment of Inquiry Skills

For the inquiry pre- and post-interviews, 24 students [N(MORE) = 12, N(Control) = 12]) were provided with a brief description of the experimental questions, methods, data, and conclusions for two studies. For the post-interview, administered during the 13th week of the semester, the two studies investigated the relationship between silicone breast implants and immunological disease. The conclusions reached in the two studies were contradictory, and thus provided clear avenues for the students to be critical in their evaluations. Students were asked to comment on the validity of the conclusion of each study (*explanation*), to critique the methodology (*critique*), and to offer evidence that would disconfirm their personal beliefs about the relationship between silicone implants and immunological disease (*evidence*). A student's total interview score comprised the sum of the scores on these three dimensions. Analyses of the post-interview responses showed that the MORE group scored significantly higher overall (unpaired t-test $p < 0.05$; Cohen's *d* effect size = 0.78, a large effect size), with the only significant difference in subcategories in the explanation dimension (unpaired t-test $p < 0.05$; Cohen's *d* effect size = 1.39, a large effect size). Taken together, the results from the experimental design skill and interview-based assessments suggest that the increased inquiry responsibility and emphasis on explanations in the MORE curriculum helped students learn to analyze data and present coherent explanations.

Example Learning Outcomes Based on Analyses of Course Examination Data

We also compared MORE (N = 39) and Control (N = 38) student performance on their regular, lecture-based course examinations. The four exams were prepared by the course professor and head teaching assistants who were not involved in the research study comparing the different laboratory experiences. The exams focused on the material covered in the professor's lectures and application of mathematical relationships, and included both multiple-choice and free-response question formats, as well as isomorphic and near-transfer type questions. Graduate student instructors scored the exams under the supervision of the professor.

As shown in Figure 1, the mean exam scores of the MORE and Control groups diverged across the semester, with the MORE students' average increasing and the Control students' average decreasing. By exams 2 and 3, the MORE students were performing marginally better than the Control students (two-tailed, unpaired t-test $p < 0.10$). At the end of the semester, the MORE group performed significantly better than the Control group, scoring 8% higher on the comprehensive final examination (two-tailed, unpaired t-test $p < 0.05$; Cohen's *d* effect size = 0.53, a medium effect size). There were no statistically-significant differences in the exam means of the Control group compared with students in all other standard laboratory sections in the course.

In addition, the MORE group significantly outperformed the Control group on several individual final exam items, while the Control group did not outperform the MORE group on any of the final exam items. The largest performance differences between the MORE and Control groups on the final examination were observed for broad topics (such as acid-base chemistry) that students investigated in both the MORE and Control laboratory curricula. For these topics, the MORE students averaged 11% higher than the Control group students ($p < 0.05$, Cohen's *d* effect size = 0.70, a large effect size). Furthermore, although there were significant differences favoring the MORE groups for both isomorphic and near-transfer exam questions, the differences were larger for the questions requiring transfer. These results are consistent with the emphasis in MORE instruction on practicing reflection and developing an understanding of chemistry ideas, as opposed to practicing algorithmic chemistry exercises.

Several post-survey items also solicited students' opinions about the examinations and the studying strategies that were most effective for performing well on them. Interestingly, while MORE and Control group students did not report any differences regarding their studying habits, a statistically-significant difference emerged in response to the question "How do you think you could have improved your exam scores?" Thirty-nine percent of MORE students and 17% of Control students reported that they believed they could have increased their exam scores by improving their depth of understanding (Fisher's exact test $p < 0.05$). Specifically, students cited a need to focus more on integration or understanding of concepts, asking questions, or explaining ideas as part of their exam preparation.

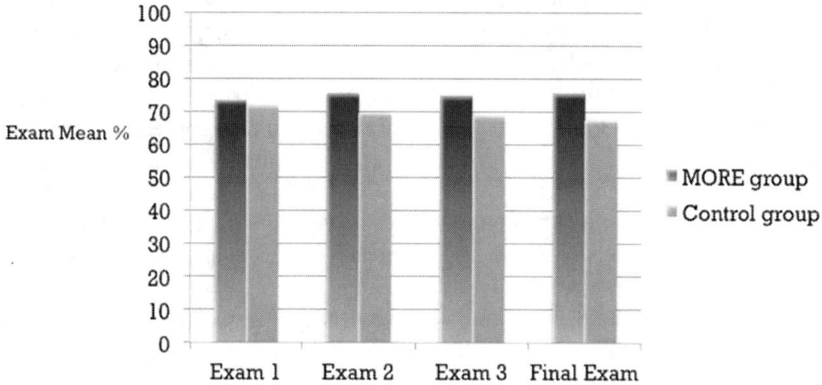

Figure 1. Comparison of exam means of MORE group and Control group across the semester. Differences between groups are marginal (two-tailed, unpaired t-test $p < 0.10$) for exam 2 and exam 3, and statistically significant (two-tailed, unpaired t-test $p < 0.05$) for the comprehensive final examination.

Overall, the results of these analyses indicate that students in the MORE laboratory sections may have developed superior understanding of chemistry ideas and problem-solving abilities throughout the semester compared with students in the standard laboratory sections.

Example Learning Outcomes Based on Problem-Solving Interview Data

In the problem-solving interviews focused on the topic of acid-base chemistry, conducted at the end of the semester, students [N(MORE) = 14, N(Control) = 14] were asked to solve and discuss problems that were designed to require deeper understanding of chemical principles than is typically necessary to answer course examination questions. Though there were no significant differences in interview problem-solving success between the MORE and Control groups, further insights about the possible sources of the differences in problem solving on examinations were gleaned from a comparison of MORE and Control students' problem-solving processes observed during the interview sessions.

The first interview question asked students to determine the pH and the pOH for a neutral aqueous solution at the normal temperature of the human body given the autoionization constant for water at that temperature. Students were then asked to draw pictures of what neutral solutions would look like on the molecular level at body temperature and at room temperature, and to relate those pictures to the corresponding pH and pOH values. Because the students had not been asked to consider pOHs or molecular-level views of such situations in their lecture or laboratory coursework, this interview question was categorized as a near-transfer problem.

Each student's initial response to the interview question was first characterized as containing non-contradictory correct ideas, non-contradictory incorrect ideas, or contradictory incorrect ideas. There were no significant differences between the groups' initial responses either in terms of correctness or consistency. Fifty-four percent of the students interviewed displayed contradictory ideas [N(Control) = 9, N(MORE) = 6] while working on the problem. For students whose initial ideas were contradictory, the students' reactions to these contradictory ideas were categorized. Table III summarizes these student responses, showing the percentages of MORE and Control students with initially contradictory ideas who (1) noticed their own contradictory ideas without prompting, (2) actively attempted to understand the problem (either after noticing contradictory ideas on their own or being prompted by the interviewer to consider them), and (3) succeeded at reconciling contradictory ideas.

As shown in the table, while the MORE students engaged in each of these behaviors with greater frequency than the Control students did, the only statistically-significant difference is in the proportion of students with contradictory ideas who attempted to revise their understanding of the problem once they realized that something did not make sense (Fisher's exact test $p < 0.05$). The MORE students reacted to the presence of contradictions in their own ideas by reflecting on those ideas and attempting to revise their ideas, while the Control students were more likely to acknowledge and accept their contradictory ideas without any attempts to reconcile them. For this sample, the MORE students' greater tendency to attempt to reconcile their contradictory ideas is consistent with these students' greater appreciation of the importance of model refinement in research compared with the Control group revealed by the previously-discussed survey results.

Table III. Comparison of MORE and Control students' responses to contradictory ideas during post-interview

Control	*MORE*	*Response to Contradictory Ideas*
11.1%	50.0%	noticed contradictory ideas without prompting
22.2%	83.3%	actively attempted to understand the problem
11.1%	50.0%	succeeded at reconciling ideas

In summary, for our original studies, we developed the MORE laboratory course, incorporating the intended instructional design principles of promoting metacognition, supporting guided discovery, and engaging students in exploration of concepts through authentic scientific inquiry. We then implemented MORE in two laboratory sections, and compared the aggregate learning outcomes of MORE students with those of students enrolled in matched Control laboratory sections. The results of our analyses indicated that we were successful in implementing the three elements of our instructional design. And, in terms of learning outcomes, MORE students exhibited significantly enhanced perceptions of learning and scientific research, inquiry skills, conceptual understanding, and problem solving abilities, particularly for problems involving the application of ideas in new contexts. In addition, based on analyses not described in detail here, we observed that MORE students exhibited significantly enhanced propensity to engage in reflection and metacognition compared with Control students (*1*, *2*).

Thus, the findings of our original design-based research provided a coherent picture of the benefits of the MORE laboratory course for student learning, as well as hints about the relationships between metacognition, understanding of ideas, and problem solving in the context of learning chemistry. Improved metacognition, in particular MORE students' greater tendency to reflect upon and subsequently revise their understanding of chemistry ideas compared with the Control students, appeared to be at least one mechanism by which the MORE students' understanding of chemistry ideas was enhanced. This was consistent with previous research in cognitive science and education that implicated metacognition as a key to developing robust understandings of ideas (*14–16*).

Individual MORE Student Cognition and Thinking Processes That Facilitate Transfer

Although randomized controlled trials are considered by some to be the "gold standard" of educational research (particularly regarding curricula and instructional methods), after carrying out our original comparison studies, we were convinced that it would be far more informative to focus our efforts on deeper understanding of the student thinking processes that led to enhanced learning outcomes for MORE students. In general, when such understanding of effective instruction is achieved, the science education community will be empowered to move beyond efforts to propagate the specific curricula and instructional methods that win the "horse races" (when compared with traditional methods). This will enable dissemination and implementation of more general, and thus more flexible, principles for effective instructional design and engaging students in key thinking processes.

Therefore, since our original MORE studies, our research has focused on refining our understanding of how students learn chemistry with a depth of understanding that facilitates success at applying models in new contexts (transfer). As part of this broad research agenda, we seek to connect an important learning outcome of our original MORE studies (enhanced transfer success, on average, for students participating in MORE) to individual student thinking processes that contribute to that outcome. This, of course, entails understanding and measuring both transfer success and the knowledge and thinking processes that may contribute to it. We have pursued this by studying student learning via the MORE laboratory module entitled "What happens when substances are added to water?" (*21*), also known as the "dissolution module". (Note that this module was developed subsequent to the original MORE studies described earlier in the chapter.) We summarize some of our progress in the following paragraphs.

First, we examined how students' molecular-level models of salt and sugar solutions evolved throughout the dissolution module, including characterizing the consistency of their models with experimental evidence, progression of models toward scientific accuracy, and strict correctness of the students' models (*22*). We investigated the effectiveness of the first implementations of the dissolution module in prompting three different populations of general chemistry students (honors students at a research university, chemistry majors at a primarily undergraduate institution, and students at a community college, total N = 84) to revise their molecular-level ideas regarding substances added to water. Understanding what happens from a molecular-level perspective when ionic and molecular compounds are dissolved in water is a foundational topic in general chemistry, yet only 15% of the students who participated in this study, all of whom had taken at least one previous chemistry course, presented correct initial ideas regarding both salt and sugar dissolved in water. Some common misconceptions found in the initial models include salt existing as "NaCl molecules" in solution, salt breaking up into neutral atoms, sugar molecules dissociating into atoms or ions, salt and/or sugar forming covalent bonds with water, and salt and/or sugar undergoing other reactions (e.g., metathesis) with water. Participation in the laboratory module led the majority of students to scientifically correct ideas in their final refined models (80% for NaCl, 52% for $C_{12}H_{22}O_{11}$), and an even greater number of students presented final models that were fully consistent with their data (89% for NaCl, 83% for $C_{12}H_{22}O_{11}$). (Note that the scientifically-correct models are not presented to students during the guided-discovery portion of instruction. Students construct their refined models based on laboratory observations and evidence. Thus, MORE instruction emphasizes consistency with experimental evidence rather than strict scientific correctness.) The results indicated that the module was particularly effective for encouraging students to revise their ideas about aqueous salt solutions such that they were both consistent with experimental data and scientifically correct, and also prompted students to make significant productive revisions to their ideas about aqueous sugar solutions.

In a subsequent study exploring students' abilities to apply the molecular-level models they developed during the dissolution module in new contexts, we interviewed students who had completed this module at a research university and a community college (*23*). Participants were interviewed at the end of the semester to investigate their abilities to apply the molecular-level models they constructed during the dissolution module in the unfamiliar context of colligative properties. During the interview, we asked participants to (1) represent their molecular-level models of various aqueous solutions, including NaCl(*aq*), in a conductivity context; and (2) read an explanatory paragraph discussing boiling point elevation, predict relative boiling points of equimolar solutions of $C_6H_{12}O_6$(*aq*) and NaCl(*aq*), and represent their molecular-level models of these solutions. We were not surprised that 95% of the interviewed students were able to draw correct molecular-level representations of NaCl(*aq*) in the initial conductivity context since it was identical to their experience in the dissolution module. However, in the boiling point elevation context, only 53% of interviewed students described NaCl(*aq*) as separated ions, a statistically-significant decrease (two-tailed Fisher's exact $p < 0.05$), even though this question was asked (on average) 15 minutes after the conductivity question. This illustrates how the nature of the context can influence the activation of students' molecular-level ideas, as the familiar conductivity context activated the correct molecular-level model while the unfamiliar boiling point context did not for some students.

The distinction between students, all of whom had participated in a MORE laboratory course, in terms of their success (or lack thereof) at applying the molecular-level models developed during the dissolution module in the new context of boiling point elevation afforded the opportunity to investigate an important research question: "What are the relationships between the knowledge and thinking processes that individual students engaged in during the dissolution laboratory module and their success at applying their model of aqueous solutions appropriately in the new context during the interview (*24*)?" For 28 students at a community college, we characterized their knowledge and thinking processes during the dissolution module based on their written initial and final refined models. In addition to describing their molecular-level ideas about salt and sugar in water in their refined models, students were asked to describe the specific changes they made compared to their initial model and the evidence that prompted those changes (or to discuss how the evidence supported their initial ideas if there were no changes). We coded students' responses to capture both their metacognitive reflections about how their understanding progressed during the module and their use of evidence to support their molecular-level ideas. Our results identified three cognitive processes that some students engaged in during the dissolution module that were highly correlated with success on the transfer interview question: (1) constructing molecular-level models that were consistent with empirical evidence; (2) engaging in high-quality metacognition by accurately reflecting on how one's molecular-level ideas changed relative to one's initial ideas; and (3) identifying evidence to support model refinements. Participation in the MORE laboratory module was not sufficient to lead to transfer success; instead, engagement in these specific cognitive processes appears to be key. These results are expected to be valuable both for refining MORE instruction to maximize its effectiveness and for developing new instruction that incorporates the facilitation of these thinking processes into its design.

In summary, the results of our studies on student learning in the chemistry laboratory illustrate both our long-term research trajectory, shifting from a focus on comparing aggregate learning outcomes to investigating individual student cognition, and the benefits of combining the different types of studies to develop a more complete picture of effective laboratory instruction and student learning. Our original comparison studies provided evidence that the design and implementation of our Model-Observe-Reflect-Explain instructional design enhanced particular learning outcomes and informed the direction of subsequent, in-depth studies that revealed specific cognitive processes that could be key for enabling chemistry learners to construct robust, molecular-level models that facilitate transfer.

Acknowledgments

This material is based upon work supported by the Camille and Henry Dreyfus Foundation and by the National Science Foundation (NSF) under grant numbers DUE-9455418, DGE-0204779, REC-0208029, and DUE-0618829. In addition, this manuscript was written while author Dawn Rickey was serving at the NSF, and includes NSF support through her Independent Research and Development plan. Any opinions, findings, and conclusions or recommendations expressed in this material are those of the authors and do not necessarily reflect the views of the Camille and Henry Dreyfus Foundation or the National Science Foundation.

We also acknowledge and thank the colleagues who have contributed to aspects of the work discussed in this chapter, in particular, Angelica M. Stacy, Melonie A. Teichert, Lisa Dysleski, Seth Anthony, Barbara A. Reisner, and Adam C. Mattox.

References

1. Tien, L. T. Ph.D. Thesis, University of California, Berkeley, CA, 1998.
2. Rickey, D. Ph.D. Thesis, University of California, Berkeley, CA, 1999.
3. Tien, L. T.; Rickey, D.; Stacy, A. M. *J. Coll. Sci. Teach.* **1999**, *18*, 318–324.
4. Brown, A. L. *J. Learn. Sci.* **1992**, *2*, 141–178.
5. Brown, A. L.; Campione, J. C. In *Classroom Lessons: Integrating Cognitive Theory and Classroom Practice*; McGilly, K., Ed.; MIT Press: Cambridge, MA, 1994; pp 229–270.
6. Brown, A. L.; Palincsar, A. S. In *Knowing, Learning, and Instruction: Essays in Honor of Robert Glaser*; Resnick, L. B., Ed.; Lawrence Erlbaum Associates: Hillsdale, NJ, 1989; pp 393–451.
7. Champagne, A. B.; Klopfer, L. E.; Anderson, J. H. *Am. J. Phys.* **1980**, *48*, 1074–1079.
8. Chinn, C. A.; Brewer, W. F. *Rev. Educ. Res.* **1993**, *63*, 1–49.
9. Collins, A.; Brown, J. S.; Newman, S. E. In *Knowing, Learning, and Instruction: Essays in Honor of Robert Glaser*; Resnick, L. B., Ed.; Lawrence Erlbaum Associates: Hillsdale, NJ, 1989; pp 453–494.
10. Schoenfeld, A. H. In *Handbook for Research on Mathematics Teaching and Learning*; Grouws, D., Ed.; Macmillan: New York, 1992; pp 334–370.
11. Schoenfeld, A. H. *Cognitive Science and Mathematics Education*; Lawrence Erlbaum Associates: Hillsdale, NJ, 1987.
12. Strike, K. A.; Posner, G. J. In *Philosophy of Science, Cognitive Science, and Educational Theory and Practice*; Duschl, R. A.; Hamilton, J., Eds.; SUNY Press: Albany, NY, 1992: pp 147–176.
13. Vygotsky, L. S. *Mind in Society: The Development of Higher Psychological Processes*; Harvard University Press: Cambridge, MA, 1978.
14. Alexander, P. A.; Judy, J. E. *Rev. Educ. Res.* **1988**, *58*, 375–404.
15. Garner, R.; Alexander, P. A. *Educ. Psychol.* **1989**, *24*, 143–158.
16. White, R. T. *J. Curric. Stud.* **1992**, *24*, 153–164.
17. White, R. T.; Gunstone, R. *Probing Understanding*; The Falmer Press: London, 1992.

18. Perkins, D. N. *Educ. Leadership.* **1986**, *43*, 4–10.
19. Rickey, D.; Stacy, A. M. *J. Chem. Educ.* **2000**, *77*, 915–920.
20. Hofstein, A.; Lunetta, V. N. *Rev. Educ. Res.* **1982**, *52*, 201–217.
21. Mattox, A. C.; Reisner, B. A.; Rickey, D. *J. Chem. Educ.* **2006**, *83*, 622–624.
22. Tien, L. T.; Teichert, M. A.; Rickey, D. *J. Chem. Educ.* **2007**, *84*, 175–181.
23. Teichert, M. A.; Tien, L. T.; Anthony, S.; Rickey, D. *Int. J. Sci. Educ.* **2008**, *30*, 1095–1114.
24. Teichert, M. A.; Tien, L. T.; Dysleski, L.; Rickey, D. Cognitive processes associated with success at applying a scientific model in a new context. **2013**, submitted for publication.

Chapter 4

Twenty Years of Learning in the Cooperative General Chemistry Laboratory

Melanie M. Cooper*[,1] and Santiago Sandi-Urena[2]

[1]Department of Chemistry, Michigan State University, East Lansing, Michigan 48824
[2]Department of Chemistry, University of South Florida, 4202 E. Fowler Avenue, CHE 205, Tampa, Florida 33620
***E-mail: mmc@msu.edu**

Chemists and chemistry educators alike defend with fervor the role of the academic laboratory in learning chemistry. However, learning in the college chemistry laboratory continues to be an under-researched field and efforts to implement effective chemistry laboratory programs have been slow to emerge. This is almost certainly due to multiple factors, which may or may not include a lack of appropriate support to conduct research in this field, lack of a reward structure for curriculum development, a lack of adequate dissemination of effective approaches to lab, and perhaps an innate resistance to change. This chapter provides a narrative of the development of a cooperative chemistry laboratory program over the course of twenty years. It begins with a discussion of the chemistry education environment when the program was introduced and how this context influenced the design and the initial attempts to assess outcomes. This is followed by a closer look at recent research studies on a wide range of outcomes for both the students in the laboratories and the graduate teaching assistants. Finally it presents arguments that current laboratory programs should be guided by learning theories and build on the outcomes of robust research on learning in the chemistry laboratory.

© 2013 American Chemical Society

Introduction

In 1991, we received funding (from both NSF and FIPSE) to develop a new laboratory program for general chemistry. The grants' duration was three years and over that time the program that became "Cooperative Chemistry Laboratories" emerged. The cooperative chemistry lab program is still in place in the original site, and the laboratory manual that accompanies it will be published in its sixth edition in 2014 (*1*). Over the past 20 years cooperative chemistry laboratories have been implemented at numerous sites, both large and small, from high school to large state universities. While there is increasing evidence about the factors that make for an effective general chemistry laboratory program, changes have been slow to emerge. This is almost certainly due to multiple factors, which may or may not include a lack of appropriate reward structure for curriculum development, a lack of effective dissemination of effective approaches to lab, an innate resistance to change, or perhaps even a satisfaction with the status quo. In fact, studies on the glacial pace of change in STEM education point to all these factors (*2*). Whatever the reason, most general chemistry laboratory programs still focus on traditional "cookbook" exercises, despite the lack of evidence for their efficacy. Our goal in this chapter is to provide a narrative of the past twenty years in the cooperative chemistry laboratory program, beginning with the development of the laboratories and a discussion of the state of chemistry education environment when they were created. We will discuss our initial attempts at assessing the outcomes from these laboratories, and why our attempts were not as fruitful as they might have been. Then we will move forward, to look at more recent research studies on a wide range of outcomes for both the students in the laboratories and the graduate teaching assistants, and finally we will present arguments that future laboratory programs should build on what is known. That is they should be guided by evidence and learning theories.

In the early 1990's chemistry education research at the college level was just beginning. Although there was a large and vibrant community of chemical educators, and the first research program in chemistry education had been established at Purdue, most curriculum initiatives were designed without a theory base to support learning, nor was evidence available to either support or refute proposed changes. Personal experience and good intentions dominated curriculum development at the college level. What has now come to be known as Discipline-Based Education Research (DBER) (*3*), was in its infancy, and it was rare for a chemistry department to hire a tenure track faculty member whose scholarly efforts focused on education research. Most faculty who were interested in chemical education were not trained in educational research methods, but rather had transitioned after tenure in a traditional sub-discipline of chemistry. The Journal of Chemical Education (JCE) did not have a "research" section until 1997, and even then, Chemical Education Research (CER) papers shared space with discussions of current traditional research adapted for the readership of JCE. The Division of Chemical Education established a committee on CER in 1994, and that committee published a report on guidelines for chemical education research (*4*). Looking back we can see this period as the beginning of a more evidence-based perspective for chemistry education. All this is to say that in

1990, there was little research about how to develop effective laboratory programs (or any other type of college level chemistry program). Although there was a significant body of research on teaching and learning emerging from Colleges of Education and Departments of Psychology there was little awareness in most chemistry departments.

College level chemistry courses were designed by disciplinary experts who did not have a background in education, and who were unlikely to incorporate any of the extant research on teaching and learning into their curricula or pedagogies. However, at the time there was a growing recognition that introductory chemistry courses were not meeting students' needs, and the National Science Foundation and the Fund for the Improvement of Post Secondary Education (FIPSE) had both targeted proposals that were designed to improve student experiences.

The Cooperative Chemistry Laboratory program was developed as a response to these calls. The goal was to redesign the general chemistry laboratories that were, at that time, the antithesis of science and the way it is practiced. Students received full line-by-line instructions on what to do, filled in worksheets and performed calculations to show how close to the known answer they could get. While this approach may allow students to learn laboratory techniques, there was little evidence even then, that a traditional cookbook lab would support learning of anything else (*4*). That is the labs might be useful for training technicians – but not for educating scientists.

It should be noted here that the circumstances that led to implementation of the new program were quite unique. External funding from NSF gave us the imprimatur to try something new and frankly, for that time, quite radical. We were in a position to make changes without affecting other faculty. (One reason why the lecture component of general chemistry is perhaps even more mired in the past is that any changes must be adopted by a committee). The goal of the new lab course was to provide general chemistry students with a laboratory experience that was closer to a research experience; that is we wanted students to solve problems by designing and refining experiments, analyzing data, using that data to make claims, and supporting those claims with explanations. In our new laboratories students would have the opportunity, in the light of evidence, to redesign experimental procedures, and have time to recover from failure. Instead of filling out data sheets student would write reports and present posters on their projects, allowing them to practice and develop their communications skills. That is we wanted to provide an experience for the 1500 students who enrolled in general chemistry each year that was a more authentic experience. Although we were not able to provide an individual research experience for these high numbers of students, we wanted to approximate it as closely as possible. However, there were a number of good reasons why most laboratory programs were tightly controlled sets of exercises, where the outcomes were well documented and the students, Graduate Teaching Assistants, (GTAs) and laboratory supervisor knew what to expect.

1. Clearly each of the 1500 students could not perform their own set of experiments, since this would cause an organizational nightmare.
2. Grading worksheets is much less time demanding than grading oral and written reports.

3. Perhaps most important, what we were asking of the students was far more cognitively demanding than following procedures.

The curricular design we settled on was one in which we allowed students "constrained freedom", in which students worked in groups on projects where we posed the initial problem, and the group developed the experimental plans and data analysis techniques. Our solution was to put into place a cooperative learning environment, in which students worked in groups on a project that lasted between 2-4 weeks. At the time cooperative learning had the largest research base of any pedagogical approach (*5*), and had been shown to improve outcomes in a wide range of disciplines and age ranges. Since then much has been learned about the mechanisms by which socially mediated learning produces gains, both in learning and in affective outcome (*3*).

The changes in the laboratory program were:

1. Students work in groups, with specified roles and duties. Experimental procedures are designed and revised by the group.
2. Each student conducts part of the experimental plan, and then the students pool data for analysis.
3. Individual accountability is accomplished by requiring individual reports from students. While the results section is common to the group, students must discuss the data and construct explanations individually.
4. The number of projects per semester is reduced from 12 to three or four each extending over 2-4 weeks, thus reducing grading for TAs.
5. Students learn laboratory techniques as a means to an end in the context of the experiment.

Each project is presented in the form of a vignette, placing the problem in context (for example analyzing an unknown compound found in a landfill, or industrial espionage to determine phosphate content of soft drinks). Each week students write a group summary of that week's experiments, and a plan for the subsequent weeks' experiments (in later years guiding questions have been provided to help students think about what might need to be done next).

The Cooperative Chemistry Laboratory Program was developed and implemented over the span of the three-year funding cycle. Looking back this seems an impossibly rapid way to implement such a radical change. Projects were developed, tested and full-scale implementation had to be completed during this time frame. Indeed, a rudimentary assessment plan was also included in the original funding request, which required one year of the project in which half the laboratories in general chemistry were cooperative and half traditional so that comparisons could be made between the two groups of students.

Since its inception in 1994, we have produced ten reports directly related to the assessment of the cooperative general chemistry laboratory program. As a research field matures, the nature of the questions addressed and its methods evolve concordantly. The theoretical and methodological frameworks we have utilized in the investigations originating in these reports reflect some of the changes that CER as a scholarly field has experienced. We developed a comprehensive program that focuses on learning instead of instruction and that would shed light regarding the attainment of the goals that prompted implementation of the cooperative project-based laboratories. Rather than summarizing the evidence in the individual studies, we highlight some of the main aspects and contributions from these studies to understanding learning in the cooperative laboratory.

Early Assessment Approaches

In the early 90s, statistical comparison of control and treatment groups in a simple pseudoexperimental design was adequate to assess the implementation of new instructional methodologies in college chemistry. Typically, the metrics employed were those already in place to assess student success in the class—exam grades and retention rates—and some basic aspect of the attitudinal dimension, often student satisfaction ("likeability" of the intervention). Alignment between specific goals of the intervention and assessment were less common. In fact, robust design of the laboratory experience as we understand it currently was not prevalent.

Within this approach, data gathered during the piloting of cooperative laboratories showed a positive effect on lecture performance and retention for female students when compared with other female participants enrolled in the conventional laboratory (*6*). No significant effect was observed for male participants. The comparative nature of this approach limited the longitudinal assessment of the format since access to control groups was impossible once the conventional program was fully phased out. In fact the short-term nature of the funding structures precluded any further comparison between control and treatment groups.

As qualitative inquiry gained recognition and grew in importance in chemical education research at the turn of the century, methods to assess the cooperative lab program also changed. The idea that simple manipulation of a variable in a laboratory setting might produce significant effects, when measured by traditional assessments such as multiple-choice tests was clearly quite naive. The assumption that the value of the laboratory experience was quantifiable in a traditional sense may not be appropriate in many cases.

In an initial attempt to incorporate these new insights, we developed a qualitative assessment plan to investigate the effect of the cooperative format in the reformed organic laboratory program (*7*). Although different from the initial general chemistry program, the organic laboratories served as platform to test research methods and to replicate previous findings. The qualitative investigation protocol was carried out during the piloting phase of the implementation and consisted of videotaped laboratory sessions, and student interviews in the case of the cooperative lab participants. An open-ended survey was administered to students in both conditions. The goal of this study was to assess the effects of the laboratory course on students' attitudes and perceptions. Although the information gathered was different in nature from the quantitative assessment utilized previously, findings were consistent. Moreover, beyond looking at simplified measurable outcomes, this approach recognized the complexity of learning in experimental settings and was a first attempt to investigate the learning processes taking place during practical activities and some of the wide variety of environmental factors that impact learning.

This investigation offered preliminary insight into the role of group dynamics, and social norms in this chemistry laboratory. It highlighted the centrality of the laboratory instructor's competence in facilitating the cooperative lab as perceived by the students. In addition, this research produced initial evidence of students' ability to clearly identify and understand the key characteristics and goals of cooperative learning as part of a lived experience and not as decontextualized knowledge. This is to say this understanding was *experiential* rather than *declarative* knowledge gained through direct instruction about the nature of the lab. Perhaps most significantly, cooperative students reported a sense of agency in their learning; however, this aspect was not fully investigated then. Participants reported experiential understanding that their role in the lab comprised "figuring out why am I doing this" and "to understand (*7*)" the experiments as opposed to conventional style laboratory students who "for the most part had a very passive view of what they were in the lab to do: basically to listen, watch and learn (*7*)". In alignment with these views, cooperative students were more prone to describe the teaching assistant as someone who would guide but not hand out answers or solutions whereas conventional students reported viewing the GTA as a supervisor, someone in charge of teaching and orchestrating what was happening in the laboratory. Interestingly, despite differences in methodological framing, studies performed years later showed remarkably similar findings.

Quantifying Learning in the Cooperative Lab: Development of Instruments

The common methods of laboratory assessment—such as perceived gains gathered through Likert scale questionnaires—do not necessarily measure the important science learning goals attainable through laboratory instruction (*8*). Laboratory represents a significantly different area of science learning than that of content acquisition associated typically with lecture (*9*). However, this fundamental realization has not translated into the assessment of laboratory effectiveness. To improve research on learning in the laboratory new research designs and methodologies are required that addressed those unique goals (Lazarowitz and Tamir, 1994, as cited by Nahkleh (*9*)). In 2005 we embarked on an in-depth investigation of the learning environment using various methodological approaches and perspectives to capture some of the elusive evidence for the relationship between laboratory activities and learning in college chemistry to which Hofstein and Lunetta had referred (*10*, *11*). We believe that metacognitive skillfulness is an intrinsic part of science practices and literature supports its tight connection with problem solving ability (*13–16*). Therefore, we initially set out to develop an across method-and-time assessment of metacognition in chemistry problem solving to examine the effect of the laboratory program. This multi-method approach involved two instruments. 1) The Metacognitive Activities Inventory, MCAi, a self report that was completed before problem solving, and was developed as part of this project (*12*), and 2) IMMEX a concurrent, automated and interactive web-based problem solving platform that can cluster and model student solution patterns and information into strategies describing metacognitive levels (*13–15*). Our studies supported the convergence of these two instruments (*13*) and provided evidence for the validity of the multi-method approach. That is, we could detect changes in student problem solving approach and metacognitive activity after an intervention (*16*).

The access to a reliable, efficient, multi-method approach was of great significance in our progress towards the assessment of the laboratory program. It allowed rapid collection of relevant empirical data after students had participated in the laboratory experience to determine its effect on metacognition and problem solving skills (*17*). In turn, this assessment approach opened opportunities for the use of more sophisticated quantitative study designs (e.g. use of Solomon Four-Group Design), processing of large cohorts of participants, and performance of multiple replication studies. This approach also allowed investigation of learning outcomes different from concept acquisition and that were not subordinate to lecture. Consistent results from three replications produced quantitative evidence suggesting that the cooperative project-based laboratory program as implemented improved students' ability to solve ill-structured problems and the metacognitive level of their solution strategies (*17*, *18*). Findings from this quantitative work were also supported by other researchers who have used qualitative inquiry methods to probe metacognition in chemistry laboratory environments rich in social interaction (*19–21*).

Mixed-Methods: Insight into Students' Lived Experience To Explain Benefits in the Cooperative Lab

Interesting as these findings were, we were cognizant of the lack of explanatory power of solely quantitative approaches. We therefore implemented a mixed methods sequential explanatory (*22*), quantitatively driven design to further investigate students experiences in the academic general chemistry laboratory. This approach involves gathering quantitative evidence to explore changes that occurred as consequence of the intervention (i.e. what occurred) and then collecting qualitative data with the intention of explaining how and why those changes occurred (*22*). We decided to investigate the laboratory experience as lived by the participants, that is, through their participant observer lens. Initially, the participants were the students but as our work evolved, we extended the use of this methodology to include the graduate teaching assistants as participants. We used a phenomenological approach as the qualitative component of our design because it could provide access to understanding of "the meaning of a chosen human experience by describing the lived experience or phenomenon as perceived by the participants (*23*)". Previously, Casey (*23*) suggested the potential of phenomenology (*24*, *25*) as a research tool to study the academic chemistry laboratory experience; however, to the best of our knowledge it had not been utilized for this purpose before. In all of our phenomenological work, data collection utilized a semi-structured interview protocol designed to promote participants' reflection about and reconstruction of their laboratory experience. A significant characteristic of this approach is that it provides the participants with an opportunity to introduce and discuss topics of interest to them and reflect those aspects that are truly essential to their experience.

In this first phenomenological study, deep textual analysis of the interviews led to the construction of an outcome space, Figure 1, that describes students' experience in terms of three fundamental dimensions: Affective Response, Understanding the Experience and Strategic Response. In essence, students find themselves immersed in a learning environment with unexpected demands that trigger affective conflict and cognitive imbalance. With 'nowhere to go', students engage in understanding the laboratory operative level and its expectations. This understanding is developed by participating and contributing in the establishment of the laboratory culture, and this constitutes the initial step in their regaining control. As they become empowered by their gradual increase in understanding of the lab paradigm, students start implementing and/or further developing the skills needed to complete the task successfully. They participate in adaptive learning behavior such as sharing and refuting ideas, asking questions and providing feedback, attempting to learn, planning, evaluating, sorting information and doing new things as required by the lab format. This skillfulness dimension brings forth evidence indicating that as a result of the demands in *this* laboratory environment the students engage actively and deeply in metacognitive behavior. "Figuring out" becomes a social norm within the laboratory experience that encourages engagement in reflection and argumentation, and promotes feedback and reciprocal explaining.

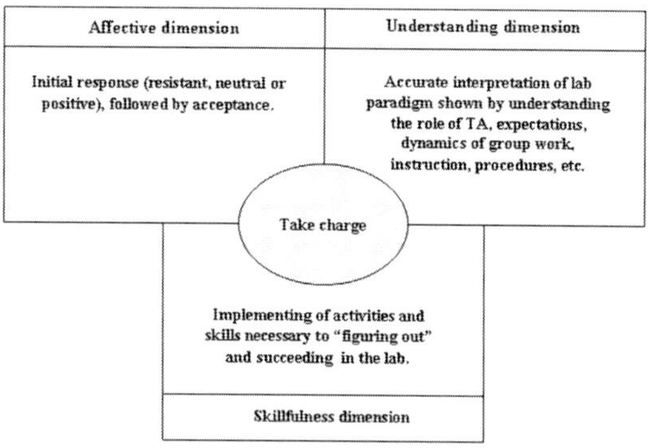

Figure 1. Outcome space for the experience of students in a cooperative laboratory. Reproduced with permission from reference (35). Copyright 2011 The Royal Society of Chemistry.

We contend that it is in this core dimension where the students develop the metacognitive and problem solving skills that we detected quantitatively using the multi-methods approach. We argue that taking charge is the interconnecting factor that brings cohesiveness to the experience. Initially, learning in the cooperative laboratories is facilitated for the students but ultimately it becomes their responsibility. Taking charge of their learning is the overarching requisite of success and students' actions and decisions need to be concerted in that direction. This directly addresses Gunstone's contention that the challenge is "to help learners take control of their own learning in the search for understanding" as cited by Hofstein and Lunetta (*26*).

Two fundamental components of this type of learning experience are the intense purposeful social interaction—which must be clearly differentiated from the activity for the activity's sake—and an environment that is conducive to the exercise of metacognitive skillfulness. We postulate that the combination of these factors significantly promotes the development of problem solving skills and learning. To complement and extend this study, we are currently in the process of analyzing phenomenological data gathered from an independent traditional, verification-oriented laboratory setting. Likewise, we have started another complementary study that involves students who experienced both, a traditional and a reformed general chemistry laboratory in consecutive semesters.

Learning from Teaching: Facilitating Academic Labs as Part of the Graduate Experience and Scientific Development

One of the outcomes of our research program is that it generates new questions, and suggests new methods to address them. For example, the pivotal role of the Graduate Teaching Assistant (GTA) in the development of an environment that led to improvements in metacognitive and problem solving skills led us to hypothesize that their engagement in modeling and promoting the desired behaviors made them active participants in the learning environment. It is likely then that GTAs participation in the laboratory affects not only the undergraduate students, but the GTAs themselves.

Up until our design of this study, most of the research on the academic chemistry lab had focused exclusively on the implementation of different instructional approaches, and students' gains and perceptions (*11*, *27*). Interest in teaching assistants had revolved around training for given instructional paradigms (*28–31*), and their perceived expectations (*32*, *33*). The sparse chemistry-related literature reported GTA gains mostly limited to content mastery, teaching ability (most likely understood as ability to deliver information), and GTAs' satisfaction. In terms of methodology, these reports were typically based on data obtained through surveys rather than resulting from a thorough analysis of the learning environment. In 2002 French and Russell reported (via a standard self report) that GTAs in an inquiry-based introductory biology course believed their involvement had indeed contributed to their ability to do research.

We decided to scrutinize the laboratory experience as lived by the GTAs, that is, through their participant-observer lens using phenomenology as the methodological framework. Our first study with GTAs included a cohort of thirteen first year graduate student volunteers with no student experience in the cooperative general chemistry laboratory (*34*). Systematic data analysis and interpretation (*24*, *35*) produced an outcome space (Figure 2) characterized by three core dimensions: the Affective Engagement, the Metacognitive Engagement and the Epistemological Reflection. The last two dimensions, the Metacognitive Engagement and the Epistemological Reflection, informed us directly about gains in intellectual skills. Meanwhile, the Affective Engagement dimension contributes in sense making and brings unity and cohesion to the model.

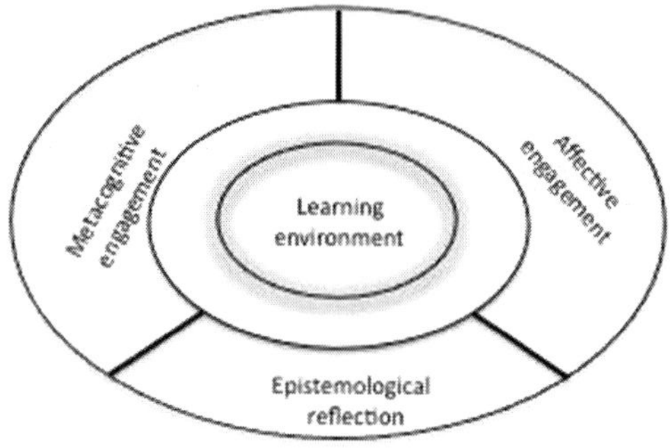

Figure 2. Outcome space for the experience of GTAs in a cooperative laboratory. Reproduced with permission from reference (34). Copyright 2011 The Royal Society of Chemistry.

There are several similarities between the GTAs experiences and those of their students in this cooperative lab program. For instance, despite induction training into their functions, the GTAs experienced uncertainty and some degree of confusion when they first faced the learning environment. Anxiety and feelings of unpreparedness are not uncommon even after longer periods of training (*28*). The GTAs had theoretical understanding of where they stood and of their goals; nevertheless, translating them into action in a setting that was unfamiliar continued to be a major challenge. "Not knowing what to do" constitutes a real problem (*36*) when contrasted with navigating a familiar and predictable environment. The latter is the case where students follow step-by-step instructions that are reinforced by mini-lectures and where provisions can be made to prevent student failures. For the sake of clarification, this "not knowing what to do" should not be equated with unpreparedness or disinterest. We contend that in responding to the demands of the environment, GTAs activated their problem solving and implemented a series of skills and strategies they found necessary to succeed in accomplishing their goals and fulfilling their responsibilities. In other words, the environment made it necessary to develop and practice metacognitive reflection and skills. Constantly planning, monitoring and evaluating their performance in conducting the task of interest became a staple in their lab work in and out of the laboratory room. Similar to the case of their students, this response to the environment's demands defined their experience.

Sudden introduction to a new and unfamiliar learning environment triggered an epistemological conflict: GTAs had to think about their role and function in instruction. They had to decipher the meaning of knowledge and learning in the new environment and to contrast it with their previous ideas. The reflection triggered by this epistemological conflict may be a prerequisite to developing a more sophisticated philosophical stance (*37*). This effect was apparently more pronounced for the GTAs than for the students.

In summary, we gathered evidence that showed that the learning environment promoted changes in GTAs ideas about teaching and learning. This study also produced evidence of GTAs implementation, practice and development of metacognitive strategies as a means to fulfill their goals in a situation that mimics some aspects of their future research tasks. The significant role of metacognition in research is well established (*13*, *16*). In addition, beliefs about knowledge and learning are strongly tied to identity development (*38*), and shape expectations and ways of learning. Therefore, we contend that appropriate teaching experiences can prepare graduate students in their journey towards becoming independent scientists and researchers. Participation in instruction has been considered an important tool to improve teaching and communication skills; however we propose that it be reconceptualized as an integral component of the graduate experience itself.

The way that graduate students viewed themselves in their role as instructors, that is, their GTA self-image, stood out within the Affective Dimension in this study. Consistently, cooperative GTAs thought of themselves as mentors rather than as 'knowledge providers' or 'managers'. It became evident that the GTA self-image the graduate students constructed was a significant factor not only influencing their own experience but it also affected the learning environment experienced by their students. This observation led to another research question: What factors influence graduate students' construction of their GTA self-image? Before addressing this question, we decided to investigate the experience of GTAs in a conventional, verification-based laboratory program with the intention of elucidating whether the outcomes we had observed in the cooperative program were intrinsic to laboratory teaching and the experience of being a new graduate student or related closely to the instructional approach.

We undertook a phenomenological study to explore the meaning that eleven GTAs ascribed to their teaching experience in a traditional, verification-based laboratory program (*39*). Phenomenological reduction and analysis of interviews produced three core dimensions that described the experience: Doing, Knowing, and Transferring, Figure 3. The perceived GTA role emerged as the interconnecting factor among them. Although this study supported the notion that GTA self-image shaped their instructional decisions regarding the learning environment, in contrast with the cooperative labs, in the verification program the GTAs viewed themselves as providers of knowledge and managers of time and safety. In their view, they were indispensable for the functioning of the lab and for their students completing their tasks. Moreover, the gains accessible to these GTAs were related exclusively to content mastery, communication skills, and personal satisfaction. Gains related to the development of more sophisticated views of knowledge and science did not emerge.

The description of the GTAs experience in this case was static in the sense that snapshots taken at different times during the semester-long experience were mostly indistinguishable. By this we mean that a progression in the nature or qualities of the lived experience were not spontaneously elicited when these GTAs were interviewed. That is they did not make explicit or implicit references to changes over the course of the experience. Trajectories of change are associated with opportunities for learning and reflection, as we observed in the cooperative laboratory program where GTAs' beliefs about learning were challenged (*35*). An apparent norm in this verification environment was to strive to create an experience that was impervious to student errors or incompetence. Clearly spelled out experimental guidelines and direct instruction were used to create a foolproof experience. Although obstacles occurred (e.g. equipment malfunctioning and shortage of chemicals) and troubleshooting was necessary, these were not problems creating a cognitive or affective conflict of the same nature experienced by GTAs in the cooperative laboratory. Problems were merely obstacles that hindered the smooth running of the scripted experience.

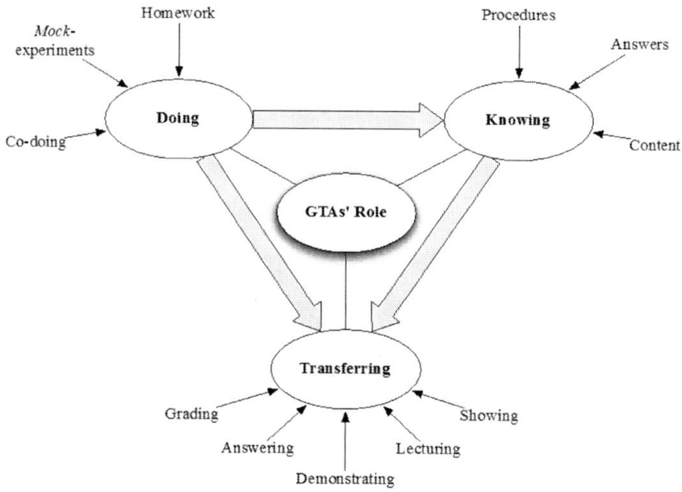

Figure 3. Outcome space for the experience of GTAs in a verification laboratory. Reproduced with permission from reference (39). Copyright 2012 Universidad Nacional Autónoma de México.

Through our in-depth, separate analyses of these two GTA experiences, it became apparent that despite the dissimilar nature of the environments, there were notable parallels not in the outcome but in a fundamental component of the experience. As mentioned above, the construction of the GTA self-image was a determining factor in GTAs decision-making and implementation of the learning experience for their students. We therefore set out to conduct an embedded, multiple case design study to gain understanding of the processes associated with GTAs' construction of their self-image as instructors and the factors influencing these processes (*40*). We believe that understanding how the GTAs self-mage is constructed would result in an increase in fidelity of the enactment of the designed curriculum and would also assist in procuring a more fruitful experience for the GTAs. In an embedded multiple case study, researchers conduct independent case studies and make comparisons within each case and across the cases. The cases in this study are the experiences of GTAs in the two diverse General Chemistry Laboratory programs and the units of analysis are the GTAs within each case.

Semi-structured interview protocols served as primary data collection for both cases and their analyses followed the explanation-building strategy (*41*). Data analysis produced a model that summarizes the factors that we propose are associated with GTA self-image development, Figure 4.

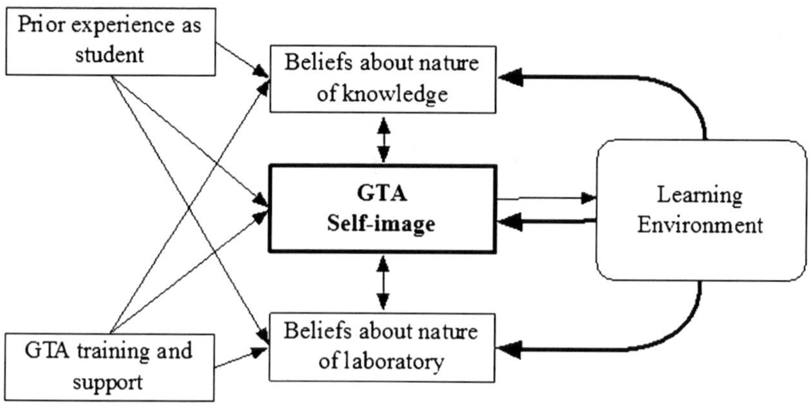

Figure 4. Factors associated with GTA self-image development.

Figure 4 summarizes the propositions supported by this multi-case study. Our evidence suggests that the construction of GTA self-image is associated with five factors: prior experiences, training, beliefs about the nature of knowledge and laboratory work, and teaching experiences. Across the two cases, we observed vastly different GTA self-images. We propose that these were not present at the time grad students entered into their respective programs but rather that they derived from the presence, or lack thereof, of conceptual, epistemological and affective conflicts and from the ways the GTAs framed their training and teaching experience once they started their assignment. To further substantiate this argument, we are currently conducting the case-study analysis of mixed-methods data that we collected from a cohort of cooperative lab GTAs at two different times. We have data from the same graduate students before they underwent any induction training and started their teaching position and after they had completed a full semester as GTAs, and will analyze these to identify shifts in beliefs.

We believe our self-image model (Figure 4) can assist laboratory coordinators as they re-consider GTA training and continuous support in a new and different light. Mechanisms of intervention and modifications may target aspects within the program that can lead to a GTA self-image more in accord with the specific instructional objectives. For example, GTAs' incoming beliefs and how those beliefs may lead them to frame what they are being asked to do could become a focal point of training programs (*42*, *43*) particularly when implementing reformed-based programs (*30*, *42*, *44–46*). GTA training programs rarely consider the impact that teaching has on GTAs. This model calls specific attention to the interaction of GTAs' self-image and their experiences in the laboratory environment. Unfortunately, it is a truism that the designed curriculum and the enacted curriculum are not one and the same; a good instructional design can fail to accomplish desired goals if crucial stakeholders are not invested (*44*).

The construction of chemistry GTAs' self-image clearly deserves a place at the forefront of general chemistry laboratory reform. It has been said that the instructor is the most important factor to reach successful implementation of a lab program (*33*), but telling GTAs *what* to teach and *how* to teach it may not be enough. Our model suggests the need to operate at the level of the GTAs beliefs to assist them in enacting the desired curriculum.

Program Reform: A Case of Research Informed Curricular Re-Structuring

Over ten years ago, Gabel pointed out that the failure of laboratory experiences to promote learning is not intrinsic to the discipline but a mere consequence of the way those experiences are structured (*47*). Similarly, Johnstone and Al-Shuaili's (*48*) attributed "missing much of the point of what undergraduate laboratories have the potential to achieve" to the pressure of having to efficiently cover a schedule of laboratory activities and observed that "worksheets and blow-by-blow manuals are still alive and healthy". The fundamental issue with college chemistry laboratory instruction may be rooted in the design of the learning experience. We contend that the laboratory experience should be constructed using coherent design and theoretical frameworks, and guided by educational research findings to emphasize what can be done, learned, and improved in a laboratory setting. We concur with the recently published DBER report (*3*) that maintains that it is through well-designed laboratories that we can "help students to develop competence with scientific practices such as experimental design; argumentation; formulation of scientific questions; and use of discipline-specific equipment (*3*)".

The findings and expertise gained from this work has prompted and supports the curricular reform of a large enrollment, verification-based, general chemistry program. This represents a unique research opportunity to replicate the findings, and assessment techniques that we have used before, and to continue the improvements of our understanding of learning in the academic laboratory. We hope that researchers and practitioners are able to apply our methodological approach to a diverse array of laboratory settings to fine tune and extend the generalizability of our findings.

Concluding Remarks

We believe that over the past twenty years our efforts in laboratory development and assessment have mirrored the development of chemistry education research and its methods. Over those years we have shown that the cooperative chemistry, project based laboratory design can realize the potential of college academic chemistry laboratories. Multi-method and mixed method assessment evidence suggests that this type of chemistry lab learning environment develops metacognition and problem solving skills, and that those skills may be transferrable to another domain. For example, students who had participated in the laboratory environment showed more metacognitive strategies when solving unrelated problems. Evidence from the phenomenological studies on the students experience in labs leads us to propose that the purposeful social interaction and intense, sustained metacognitive prompting and reflection within the laboratory environment provided support and helped develop these skills.

As active participants in the learning environment, GTAs also appear to experience gains in the same dimensions as students. The catalysts of GTAs gains are essentially the same observed for students: the sustained metacognitive demands of the problem situation, the need to solve the problem and succeed in attaining the assigned tasks, and the social interaction. In the case of the GTAs, the occurrence of a conflict preceding their epistemological sophistication is more apparent than for undergraduate students. Two factors may influence this difference: a very narrow subset of first year college students become graduate students in chemistry, whereas the student population in general chemistry laboratories is much more diverse, making these two groups very different. Graduate students' views about the nature of knowledge and science may be rooted much more deeply given their experiences during their college student careers, where traditional approaches to teaching and learning are still the norm.

Our studies indicate that the GTA self-image influences (and perhaps even determines) the enactment of the planned curriculum. This self-image is in flux and can be modified by external factors, an important idea that is not usually addressed during GTA training. However, even though the learning environment is shaped by the GTA self-image, the relationship is recursive and the experience in the learning environment can modify or solidify that self-image. That is, exposure to a non-traditional learning environment where the GTA is expected to assume the role of coach, rather than leader, and where the activities are designed and implemented by students can have a significant impact. Graduate students who participate in cooperative chemistry laboratories appear to approach the instruction and laboratory experiences in a more scientifically sophisticated way.

It is clear that to move forward, methods of assessment that are specific to learning in the labs need to be developed. In our studies, the assessment of the effectiveness of the environment in developing metacognition does not rely on student evaluations or surveys directly addressing the intervention. While the affective domain is certainly important, we do not include subjective aspects such as engagement, morale, or participation as measures of laboratory effectiveness. It is not realistic to expect the manipulation of a single variable in an environment of high complexity such as learning in the chemistry lab, to produce a significant or detectable effect.

Research Focus

Understanding what promotes meaningful learning in the lab may be more important than devoting efforts to investigating instruction. A major focus of much of the extant research is focuses on finding the "recipe" for laboratory teaching. However, we know that learning in the lab is extremely contextual and complex (even more than lecture), and producing a "script" for teaching may actually be counterproductive. A more productive approach may be to first address what skills and/or knowledge are supported by learning in a laboratory environment. Here we might take note of the extensive research done at the K-12 level, reported in

"America's Lab Report (*8*)", and think about how laboratory learning can impact the "science practices" delineated in the NRC Framework for Science Education:

1. Asking questions (for science) and defining problems (for engineering)
2. Developing and using models
3. Planning and carrying out investigations
4. Analyzing and interpreting data
5. Using mathematics, information and computer technology, and computational thinking
6. Constructing explanations (for science) and designing solutions (for engineering)
7. Engaging in argument from evidence

It may very well be that addressing these practices in laboratory, and finding ways to assess the impact of learning in this way is a more productive approach to laboratory learning.

Implications and Future Directions

This chapter brings together over twenty years work on one overarching project, it mirrors the changes that have occurred in chemistry education over that time; moving from a practitioner perspective to that of fully fledged chemistry education research, from simple assessment techniques that were not particularly informative to the use of quantitative, validated multi-methods, and phenomenological research methods to elucidate both what happens in this learning environment and how. We believe this is a valuable contribution to the field, but also want to sound a note of caution. While this project began with NSF funding (of three years duration) and many of the individual studies were part of other NSF projects, this longitudinal study is highly unusual. One of the findings of the DBER report (*3*) is that there are very few longitudinal studies, and one of the strong recommendations is that more are needed. For this to happen, researchers must be able to plan many years into the future, and have confidence that such studies can be supported. At the moment this is not the case, but we can certainly advocate for more long term planning in both our research projects and to the funding agencies.

Conflicts of Interest Need To Be Minimized

There is an inherent conflict of interest when curriculum developers attempt to assess the efforts of their work, since it is very difficult to be dispassionate about ones own efforts. In our studies we made every attempt to place and maintain distance between researchers and participants during gathering of data, and have tried not to influence the workings of the learning environment; we have avoided direct instruction of any kind that may bias students' responses and performance, and/or researchers' data interpretation.

References

1. Cooper, M. M. In *Cooperative Chemistry Laboratory Manual*; McGraw-Hill: New York, 2012.
2. Henderson, C.; Beach, A.; Finkelstein, N. *J. Res. Sci. Teach.* **2011**, *48*, 952–984.
3. Singer, S. R.; Nielsen, N. R.; Schweingruber, H. A. *Discipline-Based Education Research: Understanding and Improving Learning in Undergraduate Science and Engineering*; National Academies Press: Washington, DC, 2012.
4. Bunce, D. M.; Herron, J. D.; Jones, L. *J. Chem. Educ.* **1994**, *71*, 850.
5. Johnson, D. W.; Johnson, R. T.; Smith, K. A. In *Active Learning: Cooperation in the College Classroom*; Interaction Book Company: Edina, MN, 1991.
6. Cooper, M. M. *J. Chem. Educ.* **1994**, *71*, 307.
7. Cooper, M.; Kerns, T. S. *J. Chem. Educ.* **2006**, *83*, 1356.
8. Singer, S. R.; Hilton, M. L.; Schweingruber, H. A. In *America's Lab Report: Investigations in High School Science*; National Academies Press: Washington, DC, 2006.
9. Nakhleh, M. B.; Polles, J.; Malina, E. In *Chemical Education: Towards Research-Based Practice*; Kluwer Academic Publishers: Dordrecht, The Netherlands, 2002; pp 69−94.
10. Hofstein, A.; Lunetta, V. N. *Sci. Educ.* **2004**, *88*, 28–54.
11. Hofstein, A. *Chem. Educ. Res. Pract.* **2004**, *5*, 247–264.
12. Cooper, M. M.; Sandi-Urena, S. *J. Chem. Educ.* **2009**, *86*, 240–245.
13. Cooper, M. M.; Sandi-Urena, S.; Stevens, R. H. *Chem. Educ. Res. Pract.* **2008**, *9*, 18–24.
14. Stevens, R. H.; Johnson, D. F.; Soller, A. *Cell Biol. Educ.* **2005**, *4*, 42–57.
15. Stevens, R.; Soller, A.; Cooper, M. M.; Sprang, M. In *Modeling the Development of Problem Solving Skills in Chemistry with a Web-Based Tutor*; Intelligent Tutoring Systems 7th International Conference Proceedings; 2004; pp 580−591.
16. Sandi-Urena, S.; Cooper, M. M.; Stevens, R. H. *Int. J. Sci. Educ.* **2011**, *33*, 323–340.
17. Sandi-Urena, G. *Design and Validation of a Multimethod Assessment of Metacognition and Study of the Effectiveness of Metacognitive Interventions*; ProQuest: Ann Arbor, MI, 2009.
18. Sandi-Urena, S.; Cooper, M.; Stevens, R. *J. Chem. Educ.* **2012**, *89*, 700–706.
19. Case, J.; Gunstone, R.; Lewis, A. *Res. Sci. Educ.* **2001**, *31*, 313–335.
20. Larkin, S. *Res. Sci. Educ.* **2006**, *36*, 7–27.
21. Kipnis, M.; Hofstein, A. *Inter. J. Sci. Math. Educ.* **2008**, *6*, 601–627.
22. Creswell, J. W.; Plano Clark, V. L. In *Designing and Conducting Mixed Methods Research*; Sage Publications: Thousand Oaks, CA, 2007.
23. Casey, K. In *Phenomenology*; Bodner, G. M., Orgill, M., Eds.; *Theoretical Frameworks for Research in Chemistry/Science Education*; Pearson Education, Inc.: Upper Saddle River, NJ, 2007; pp 122−131.

24. Moustakas, C. In *Phenomenological Research Methods*; Sage Publications: Thousand Oaks, CA, 1994.
25. Patton, M. Q. In *Qualitative Research & Evaluation Methods*; Sage Publications: Thousand Oaks, CA, 2002.
26. Hofstein, A.; Lunetta, V. N. *Sci. Educ.* **2004**, *88*, 28–54.
27. Domin, D. *J. Chem. Educ.* **1999**, *76*, 543–547.
28. Birk, J. P.; Kurtz, M. J. *J. Chem. Educ.* **1996**, *73*, 615–616.
29. Nurrenbern, S. C.; Mickiewicz, J. A.; Francisco, J. S. *J. Chem. Educ.* **1999**, *76*, 114–119.
30. Kurdziel, J. P.; Turner, J. A.; Luft, J. A.; Roehrig, G. H. *J. Chem. Educ.* **2003**, *80*, 1206–1210.
31. Bond-Robinson, J.; Rodriques, R. A. B. *J. Chem. Educ.* **2006**, *83*, 313.
32. French, D.; Russell, C. *BioScience* **2002**, *52*, 1036–1041.
33. Herrington, D. G.; Nakhleh, M. B. *J. Chem. Educ.* **2003**, *80*, 1205.
34. Sandi-Urena, S.; Cooper, M. M.; Gatlin, T. A. *Chem. Educ. Res. Pract.* **2011**, *12*, 92–100.
35. Sandi-Urena, S.; Cooper, M. M.; Gatlin, T. A.; Bhattacharyya, G. *Chem. Educ. Res. Pract.* **2011**, *12*, 434–442.
36. Wheatley, G. H. *MEPS Technical Report 84.01*; School Mathematics and Science, Purdue University: West Lafayette, IN, 1984.
37. Kinchin, I. M.; Hatzipanagos, S.; Turner, N. *J. Further Higher Educ.* **2009**, *33*, 45–55.
38. Baxter Magolda, M. B. *J. Coll. Stud. Dev.* **2001**, *42*, 520–534.
39. Sandi-Urena, S.; Gatlin, T. A. *Educ. Quim.* **2012**, *23* (141), 141–148.
40. Gatlin, T. A.; Sandi-Urena, S. **2013**, under review.
41. Yin, R. K. In *Case Study Research: Design and Methods*; Sage Publications, Inc.: Beverly Hills, 2009.
42. Smith, L. K.; Southerland, S. A. *J. Res. Sci. Teach.* **2007**, *44*, 396–423.
43. Goertzen, R. M.; Scherr, R. E.; Elby, A. *Phys. Rev. Spec. Top.-PH* **2010**, *6*, 020125.
44. Seymour, E. In *Partners in Innovation: Teaching Assistants in College Science Courses*; Rowman & Littlefield Publishers, Inc.: Lanham, MD, 2005.
45. Southerland, S. A.; Sowell, S.; Blanchard, M.; Granger, E. *Res. Sci. Educ.* **2011**, *41*, 299–317.
46. Luft, J. A.; Kurdziel, J. P.; Roehrig, G. H.; Turner, J. *J. Res. Sci. Teach.* **2004**, *41*, 211–233.
47. Gabel, D. *J. Chem. Educ.* **1999**, *76*, 548–554.
48. Johnstone, A. H.; Al-Shuaili, A. *Univ. Chem. Educ.* **2001**, *5*, 42–51.

Chapter 5

A Trajectory of Reform in General Chemistry for Engineering Students

Thomas A. Holme* **and Heather Caruthers**

Department of Chemistry, Iowa State University, Ames, Iowa 50011
***E-mail: taholme@iastate.edu**

This chapter considers efforts to enhance learning within the general chemisty course taken by pre-engineering students. Because this course is inherently offered as a service course, often for students in a different college from the Chemistry Department at a university, there are both constraints and opportunities related to the manner in which reform can be enacted. Efforts spanning roughly 15 years are described and an emphasis on the nature of problem-solving within the course emerges as a common theme. The issue of student motivation is also considered with pre-engineering students serving as a prototype of a type of student who doesn't inherently see the value of learning chemistry.

In some respects, there is a love/hate relationship between chemistry faculty members and the large service courses in introductory chemistry that they often teach. Most are well aware that the large number of students in these courses represent a claim on university resources for their Department, but those same students can present challenges in terms of inspiring meaningful learning. One cohort that often fills this role is the entering freshman class of pre-engineering students. There is little question that these students take a rigorous set of classes and are often quite capable students, and yet the experience of many chemistry instructors is that they find motivating "the engineers" a singular challenge. In part this situation may be attributable to a learning culture that is, in measurable ways, different from that of the basic sciences (*1*). Nonetheless, it is possible to cast

© 2013 American Chemical Society

the learning objectives in terms that are more commonly appreciated by the pre-engineering students. This chapter describes one trajectory by which this type of adaptation within the "chemistry for engineering students" course has developed.

Emphasizing Problem-Solving: Gateway Exams

While it may be a stretch to suggest that engineering students are inclined to remember detailed chemistry facts, such as when precipitates will form, there are other broader constructs that are taught within the general chemistry course that are capable of providing meaningful and hopefully transferable learning for students. Other disciplines, particularly mathematics, have confronted this same premise by enacting "gateway examinations" in calculus classes (*2*, *3*).

Essentially, the idea behind gateway exams is that certain components of a service course provide specific skills needed in later courses. Once identified these skills are assessed separately on a competency basis. Students who demonstrate competency in calculus skills like differentiation and integration receive grades that allow them to continue to courses with a calculus pre-requisite, based on the expectation that the needed skills from the course are in hand. To adapt this concept to General Chemistry requires several things. First, skills that are particularly useful to engineers and might be separately assessed must be determined. Second, questions that will assess those skills must be devised. Finally, because competency based assessment allows for retaking gateways exams, the logistics of a system must be worked out.

Unlike calculus, where specific skills needed in engineering mathematics are readily enumerated, problem-solving in chemistry tends to be more closely tied to content specifics. Nonetheless, there are problem-solving skills that are likely to be transferable to engineering contexts and these skills include (1) recognizing knowns and variables in a problem; (2) being able to determine what information is missing and needs to be looked up; (3) being able to recognize relationships between variables in a problem; (4) recognizing multiple levels of complex problems and (5) being able to represent problems with diagrams. A gateway examination system based on these identified skills was implemented in a General Chemistry course for engineering students and important insights into the nature of assessing problem-solving were derived from this study.

From a cognitive theory perspective, there are several reasons to suggest that questions that are designed to elucidate student understanding of their own problem-solving skills will have useful learning outcomes. Prior research suggests that enhanced metacognitive skills are associated with higher order learning (*4–6*), and most college courses include such learning as important goals. The gateway exam approach within general chemistry therefore was designed to explicitly assess whether or not students can identify their own problem-solving strategies for various chemistry problems. Such explicit questions tend to force metacognition.

The gateway exam scheme in general chemistry for engineering students focuses on the assessment of student problem-solving. Studies associated with conceptual understanding in chemistry (*7–9*) and physics (*10*, *11*) have demonstrated that the ability of a student to get a correct numerical answer does not necessarily demonstrate that they know the science incorporated in the problem. In part because of these studies, the gateway testing regime was designed to have students to describe how they solve problems, often ones that include incomplete information, rather than simply providing numerical answers for specific chemical queries. Thus the premise is that students who cannot elucidate their problem-solving strategies are not as likely to generalize those strategies outside of chemistry.

When it was implemented, the gateway exam concept was used in the first semester of a two-semester general chemistry course taken exclusively by pre-engineering students. This course was subsequently abandoned by the College of Engineering for a one-semester alternative, as will be noted later, but the essential premise of the gateway project is not altered by that curricular modification. To maximize the opportunity to emphasize problem-solving, two gateway exams were required during this course. The first exam is given roughly 2 weeks after the first regular exam – and after both stoichiometry and introductory energy concepts have been covered in the course. The driver for student behavior was only positive, that is, there was a grade benefit for passing two exams, but there was no sanction associated with not passing both tests. The exam consists of nine questions, all of which were free response format. To provide some insight into the way such questions are worded and graded, an example of one question from the Gateway Exam are provided in Figure 1.

This type of question can cause difficulty in grading because of the use of free responses. The precise rubric is seldom given as a student response. This type of ambiguity, however, is not particularly unique – and is encountered in any free response question. Other questions require fewer components for a correct response – but this example is representative of the type of question that is in the gateway exams. All gateway exams were graded by a team of only two graders to minimize grading errors associated with interpretations of student responses.

It is important to realize that the expected responses to questions such as these are quite different from what students have been expected to produce in their previous courses. Some students noted their discomfort with questions that did not ask for a specific answer, but rather how to proceed to get an answer.

Because the gateway exam concept includes competency based examinations (*12*, *13*), it also requires the scheduling of retake exams. Students are allowed to take similar exams until they demonstrate that they are competent in the material being assessed, in this case problem-solving skills. This requirement did impose some logistical concerns that were handled via a scheduling system similar to those currently available in many course management systems. Further details about the logistics of implementation of competency based exams will not be included because they tend to be platform dependent.

For the implementation of gateway exams, students were expected to pass two separate exams (after roughly four weeks of class and again after eight weeks), so with a cohort of 93 students, a total of 393 exams were collected and graded. Unfortunately, the grading burden associated with such a large number of exams was not evenly distributed, as students clumped together when they took retake exams – mostly at times near the deadline. In addition, establishing the cutoff point for competency based testing is not without controversy (*14*). For gateway exams in general chemistry two cutoffs were established. For the first instance of the exam, taken during class time, a passing grade is 78% (70/90) while for retakes the cutoff was increased slightly to 83% (75/90). The separate levels were instituted to provide some impetus for students to try to pass on the first attempt. Even with the lower cutoff, few students (less than 5%) passed the exams on the first try. Despite the fact that few students passed the Gateway Exam on the first try – most students who made an effort to take retake exams did eventually demonstrate competency. Patterns of student retake behavior reveal some interesting trends.

> *"Copper oxide, CuO is responsible for the green color sometimes observed on copper fixtures used on houses or other buildings. CuO can be converted back into copper by a reaction with hydrogen at high temperatures,* $CuO(s) + H_2(g) \rightarrow Cu(s) + H_2O(g)$*. Suppose you have a piece of a façade from a famous building and you wish to remove the oxide coating by this reaction. If the surface area of the piece is 480* cm^2 *and the oxide coating is 1.5 mm thick, how would you determine the minimum amount of hydrogen gas you would need to carry out this particular reaction? Would you need to look anything up to solve this problem?"*

Rubric of expectations for a correct response components for the example question
1. Determine the volume of CuO from the given dimensions (conversion of mm to cm)
2. Look up the density of CuO
3. Determine the mass of CuO by multiplying density by volume
4. Look up or calculate the molar mass of CuO
5. Convert mass of CuO to moles of CuO
6. Use chemical equation to note that moles of CuO and H_2 are the same
7. From the moles of H_2 determine the mass of H_2 using the molar mass.

Figure 1. An example question and scoring rubric from the first Gateway Exam.

The most interesting pattern associated with gateway exams is the difference in the recollection of students who passed both exams versus those who did not. In an end of course survey students were asked, among other things, to identify how many times they retook gateway exams in order to demonstrate competency. These self-report results can be compared to actual student retake numbers and this data is summarized for students who did not pass two exams in Figure 2.

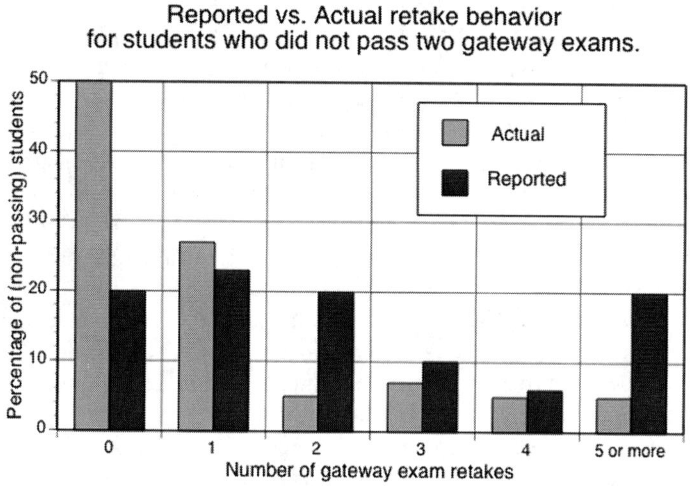

Figure 2. Actual student retake behavior compared with reported retake behavior for students who did not pass both gateway exams.

Looking at this information, which shows only students who did not pass both gateway exams, half of them did not avail themselves of retaking either exam (50% actual retakes are zero). Only 20% of these students reported that they took no gateway exams. At the other end of the spectrum, only ~5% of the studented who didn't pass both exams took 5 or more gateway exams, and yet 20% report having done so. Perhaps it is not surprising that students who struggle with the gateway exams mis-remember their efforts. Nonetheless, this evidence suggests that followup communications with students who have not yet passed gateway exams may be an important component of promoting student success.

Externally Imposed Curricular Changes

At roughly the same time that the gateway exam project was implemented, faculty and adminstrators in many colleges of engineering were confronting a realization that they had to respond to an overcrowded curriculum. Basic science courses were investigated relative to the overall credit load and in many schools, for many engineering majors, the chemistry requirement was reduced from two semesters to one. In principle, chemistry departments could respond by having students take the first semester of the two-semester sequence, and it's apparent that this response has occurred in a number of schools. From the perspective of course content, however, this choice is less than ideal because some of the topics in the second semester, such as corrosion and electrochemistry, would appear to be useful for future engineers.

At the same time, attempting to simply squeeze all the content of a full year course into a single semester is also an unattractive option. Thus, efforts to establish a sensible one-semester alternative were undertaken and implemented. Initial activities involved working with faculty from the engineering departments to identify chemistry content they viewed as particularly important. Next, general chemistry instructors were queried to identify fundamentals that could not be abandoned if the more applied topics often mentioned by the engineers were to be adequately treated. Then, for a series of four semesters, the new one-semester course was offered and measures of learning included the use of a full-year ACS general chemistry exam (*15*) to make comparisons between local students in the one-semester course and national samples of students in full-year courses. During these four semesters, course average grades on the ACS "brief" exam (selected because it only used half of the final exam testing period) were within half of a standard deviation of the national average, three times slightly below and one time slightly above. Thus, even though the one-semester course necessarily abbreviates the coverage of topics, students in the course show reasonable content knowledge when compared to a national sample of students who have taken a full, two-semester version of general chemistry.

The course thus designed includes coverage of most of the topics of the full year general chemistry course. In many cases, however, the depth of the coverage is sacrificed. Thus, topics covered include:

- Introductory concepts
- Molecules, reactions and chemical equations
- Stoichiometry
- Gases and gas laws
- Atomic structure and the periodic table
- Chemical bonding
- Materials
- Thermochemistry
- Thermodynamics
- Kinetics
- Equilibrium
- Electrochemistry

In addtion to the fundamental chemistry, however, specific efforts to cover engineering applications were also incorporated. Thus, for example, as noted earlier, the coverage of electrochemistry was explicitly tied to understanding of corrosion. Coverage of corrosion includes galvanic and crevice corrosion, in addition to uniform corrosion that is more commonly found (often briefly) in general chemistry courses. Similarly, the treatment of stoichiometry focuses significant time on the use of fuels, not excluding other important reactions, but providing an emphasis on chemistry that a significant number of engineering students might find relevant to their future studies.

A major drawback for staging this course was the lack of a textbook that utilized this approach. A number of educators recognized this problem in designing a one-semester course, and as a result several textbooks were developed (*16–18*). The idea that the chemistry content should look essentially like any general chemistry course, but that the content be couched within applications that might be more easily seen as relevant to engineering students was the core concept of the text developed by Brown and Holme (*16*), and this book is now being revised into a third edition. The chapter coverage looks similar to the bulleted list above, with the exception of an additional chapter on nuclear chemistry. Engineering contexts that have emerged for the course include a diverse set of topics including, air pollution, biomass and biofuels, concrete fabrication and aging, nanotechnology, and trace analysis of materials. An additional emphasis on materials related chemistry results in the incorporation of ideas about polymers, for example, throughout the text as well.

Another feature of this book is that it incorporates a feature that uses many of the problems devised in the gateway exam project. When the course was moved to the one-semester format, the curricular crowding in terms of topic coverage became such that the time devoted to the gateway exam problem-solving testing was no longer feasible. In a formal sense the gateway exam project was shut down. Nonetheless, the idea of explicitly teaching the value of problem-solving and working to help students transfer problem-solving skills remains in the course and in this textbook. At the end of each chapter, there is a section referred to as "Focus on Problem-solving" and the questions there are similar (indeed in many cases identical) to those used in the gateway exam project. Thus, the lessons learned in that project continue forward even though the competency based testing concept itself is no longer in use.

Further Investigations of Problem-Solving

Regardless of the time contraints and curricular demands of the one-semester course, the premise remains that a key developmental component of the chemistry course for pre-engineering students lies in the enhancing of problem-solving skills. An important question then becomes, what is problem-solving in this context, and how do chemistry exercises help students learn problem-solving skills? Thus, having initially implemented gateway exams, then having to forego them as a result of curricular compression, a key question still required research. Specifically, how do chemistry problems become more than just exercises (*19*) for engineering students? This question was investigated in two ways; using an online problem-solving system called IMMEX (*20–24*) and via interview-based qualitative research with pre-engineering students.

One means used to determine student problem-solving behaviors used the IMMEX system. The IMMEX system was an on-line tool that required students to solve ill-defined problems that related chemical concepts to real-life contexts. Each problem set in IMMEX was designed to have a general description of the situation, and have links to all the data and background information a student might need to solve the problem. Each problem set has different examples, or clones, with different values for variables, or different compounds to identify, and the exact example given to each student was randomized. Students solve several clones for each assignment, and this repetition results in them stabilizing into a measurable problem-solving strategy (*21*). A computer system tracked data behind-the-scenes about what links the students access within a problem set and collected that information into a database. The information in that database is then analyzed using artificial neural networks, Hidden Markov Models and sorting of student learning trajectories (*20*) into quadrants that measure both effectiveness and efficiency of problem-solving.

For the purposes explored here, the key feature of these quadrants scores is that they categorize students based on the solutions they achieve and the pathways they take to get there. Those students who fall in quadrant 1 are neither efficient nor effective at that given problem; students in quadrant 2 are efficient but not effective in their problem-solving; quadrant 3 has students who are effective at finding a correct answer but not efficiently and quadrant 4 has those students who are both efficient and effective at solving the complex problem. This ordering of quadrants represents a ranking where students would ideally progress from lower values (1 & 2) towards higher values, with a score of 4 being the goal.

In the one-semester chemistry for engineering course, during a particular semester, students were assigned five different IMMEX problems covering the topics of (1) identifying elements or compounds, and states of matter (Model Madness); (2) stoichiometry (How Much to Order); (3) gas laws (Gas Laws on Planet Ardanda); (4) thermochemistry (RXN) and (5) the qualitative identification of an unknown (Hazmat). Again, it is important to place these assignments in terms of content coverage in the course, problem 1 is based on prior knowledge of basic chemical facts. Problems 2-4 are based on material that is directly covered in class. Problem 5 requires students to identify an unknown based on the results of chemical tests, a task that was not specifically covered in the course, but requires logical application of test results that utilize familiar topics. For example, the Hazmat problem uses flame tests to allow students to identify elements present, and this concept is incorporated in the instruction of atomic structure in the course. The idea of atomic spectra being useful in this way in the laboratory, however, was not explicitly covered.

The percentage of students whose problem-solving strategies stabilize into each of the four quadrants on each of these problems is shown in Figure 3.

Looking at this graph it is possible to identify important trends that suggest the pathway pre-engineering students take related to problems versus exercises in chemistry homework. First, the percentage of students in either quadrants 1 and 2 (where effectiveness in problem-solving is less) is roughly the same through the 1st three problems, drops slightly in problem 4 and rises notably for the final problem. To the extent that obtaining a correct answer is the goal of the homework assignment that uses the IMMEX system, students in these quadrants are struggling. The number of struggling students does drop in the last "familiar" topic assigment, but rises sharply when the content is less familiar. While the details are different, the importance of content familiarity is also evident with students who do succeed in solving the problems (quadrants 3 and 4). In this case, the problem-solving goal is towards higher efficiency. Looking at problems 2 through 4, the efficiency is increasing, to the point where 83% of the students are effective at obtaining the correct answer, and with half of them doing so efficiently (quadrant 4). When the last, less familiar unknown analysis, problem is assigned the efficiency of this group of (successful) students drops precipitously. Thus, for this cohort of students at least, this data suggests that a majority of pre-engineering students progress towards more efficient problem-solving strategies throughout the course as long as the problems involve chemistry content on which they have received explicit instruction. Perhaps they are able to progress to the point where what starts as a chemistry problem becomes a chemistry exercise. That progression is strongly hindered (efficiency drops significantly) when they must utilize chemical information in less familiar ways to solve the problem.

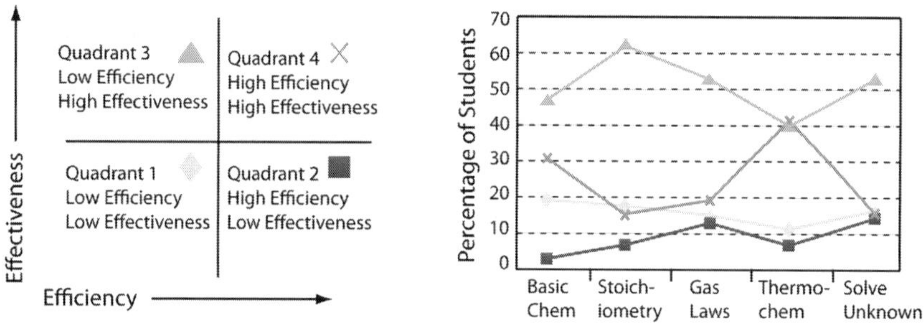

Figure 3. (left) Definitions of learning trajectory quadrants. (right) Percentage of students in each learning trajectory quadrant (number of students in sample = 650).

The second study that probed the difference between exercises and problems in general chemistry for engineering students used qualitative methods in an interview format. Twenty volunteers were solicited from a large-lecture pre-engineering general chemistry class to take part in a one and a half hour to two-hour interview at a time of mutual convenience for the interviewer and interviewee. The students were offered free food for taking part in the interview and informed consent was obtained. In order to ascertain the students' thoughts while they were working through a set of chemistry problems, a talk-aloud protocol was used during the interviews (*25*). This protocol asks students to verbalize what they are thinking about doing or why they are using a particular idea or method while they are doing it. During the interview, students had access to their textbooks, to a calculator and to the internet to be able to look up unfamiliar terms or ideas. The interviews were video and audio recorded for data collection purposes and they were transcribed as part of the data analysis process. Of the twenty volunteers, 11 students completed the two tasks analyzed here as well as having participated in a readiness diagnostic test for the course that was used to establish that similar background knowledge was present in the students whose work was analyzed. The chemistry content of the interview consisted of three questions each on stoichiometry and thermochemistry. For the purpose of the discussion here, results from the student interviews during two of the stoichiometry questions will be presented. These two questions were patterned after end-of-chapter questions in general chemistry textbooks and are presented in Table 1.

Table 1. Interview questions for stoichiometry problems

Familiar Question (exercise)	*Unfamiliar Question (problem)*
What mass of oxygen is needed to completely combust 1.00 g of ethanol to produce carbon dioxide and water?	Octane (C_8H_{18}) is a component of gasoline. Complete combustion of octane yields H_2O and CO_2. Incomplete combustion produces H_2O and CO. If 1.000 gallon of octane is burned in an engine and the total mass of CO, CO_2 and H_2O produced is 11.53 kg, what is the efficiency of the process, in other words, what fraction of the octane is converted to CO_2? The density of octane is 2.650 kg/gal.

The initial analysis of problem-solving behavior utilizes a categorization scheme proposed by Calimsiz (*26*), that identifies seven traits of problem-solving:

(1) gaining basic familiarity with the problem
(2) restating the problem
(3) searching for a starting point
(4) working from the starting point towards the final goal
(5) consulting sources
(6) modifying or abandoning a step or route
(7) evaluating the work done

While any of these behaviors can be found in a students' attempt to solve a problem, the percentage of time spent in each behavior points to the strategies being used. There is no single correct path to solving problems of this nature, but for students to reach the level of efficiency that is associated with quadrant four in the previously noted IMMEX study, an increased amount of time in productive work towards the goal (behavior 4) is probably the most likely to lead to efficient solution of the problem. A graph showing showing percentage of time spent in each of the seven problem-solving behaviors for both tasks is shown in Figure 4.

These two graphs show distinct differences that establish that students who encounter familiar style stoichiometry problems (Figure 4 - top) spend a majority of their time productively moving towards an answer (behavior 4) and looking up resources needed to achieve that goal (behavior 5). By contrast, for the unfamiliar problem that includes both complete and incomplete combustion, more exploratory behaviors are notably more common. A greater percentage of time is spent gaining familiarity with the problem and restating it (behaviors 1 and 2). While still a small percentage, more time is spent looking for a starting point (behavior 3). Relatively little time is spent looking up resources (behavior 5), though the same resources were available in both cases – including internet access. Finally, a significantly greater percentage of time is spent by students trying to evaluate their progress (behavior 7). It is certainly true that students still spend much of their time trying to "work the problem" (behavior 4), but it is also apparent that the unfamiliar problem type induces more general problem-solving behaviors. Given the challenge of this problem for most of the students, they are ultimately utilizing general strategies that are likely to be important to learn if they are to improve their overall problem-solving ability.

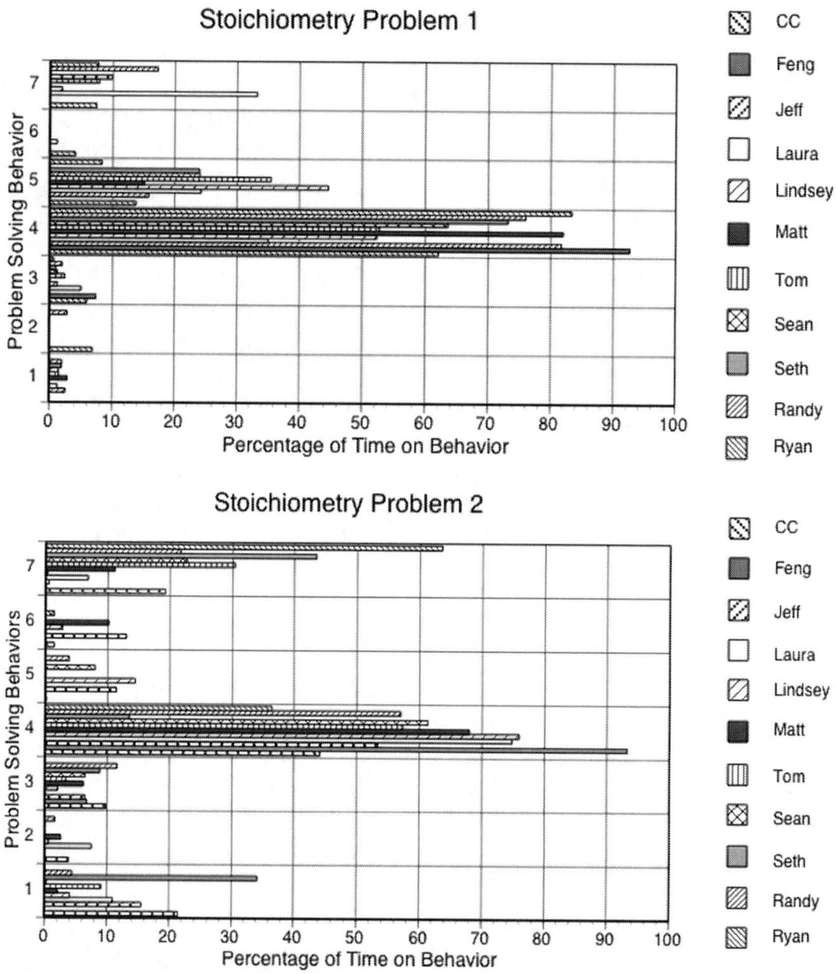

Figure 4. Problem-solving behaviors for a familiar task (top panel) and for an unfamiliar task (bottom panel).

This final observation, that it is possible to induce more problem-solving exploration with a fairly modest contextual change to a traditional stoichiometry problem is a key result. To the extent that an important learning objective for pre-engineering students in general chemistry is the development of problem-solving skills, it is noteworthy that practice of such skills can be induced with arguably modest increases in the complexity of the type of problems students are expected to do. Anectdotal supporting evidence in support of this premise arises from classwide discussions of a kinetics problem as part of homework in the course. The problem was a unimolecular dissociation into two product molecules, and the only information given was the total pressure as a function of time. In this problem, a student must be aware that the total pressure change essentially provides a second numerical relationship that allows the problem to be solved. A similar level of reasoning is needed in the octane problem used in this study.

Thus, there appear to be two ways that remove students from a problem-solving behaviors that are algorithm driven and more akin to answering exercises. As noted in the IMMEX study, changing the chemistry context appears to challenge students into using more general (and at least early on less efficient) problem-solving strategies. The second method that accomplishes this goal is to incorporate the need for students to recognize a second quantitative relationship within the problem, beyond those relationships in more "exercise" style problems.

This combination of observations has implications for how to incorporate problem-solving strategy development within the one-semester general chemistry course for engineers. The homework assignment strategies for this course have been changed to explicitly incorporate the findings presented here. Students are presented with "suggested" problems from the end-of-chapter selections that are more commonly in the exercise category so they can have the practice they may need with this level of question. These suggested problems are not handed in. Rather, a small number of the more challenging problems that are more likely to move them past exercise-level algorithmic approaches are what must be turned in for grading. Grading burdens on teaching assistants are minimized by limiting the number of these problems that are assigned. The problems occasionally include the "Focus on Problem-solving" style questions where the answers students must provide are strategies, rather than numerical answers. Students in the course are told explicitly that these assigned problems are expected to be more complex and likely more time consuming for them – and because of the time they take such problems could not be included on timed-tests, for example. Shifting such problem-solving activity to the homework side of the course is accompanied, however, by explicit discussion of strategies for approaching complex problems during lecture. Thus, while the accountability students have (points in the course) is paired with homework, the message of problem-solving strategy development is consistent in all aspects of the course.

Summary

This chapter has focused on the teaching of general chemistry for pre-engineering students. Because this course is fundamentally a service course for other majors, the premise has been taken that ways to connect to the needs of those engineering majors must be advanced. In this case, the primary transferable need that has remained the focus of attention throughout many changes in the course has been problem-solving.

Nearly any chemistry instructor would be delighted if engineering students remember a sizable portion of the chemical details of such a course, but realistic appraisal of the prospects for this outcome is not likely to be optimistic. Using problem-solving improvement as an explicit goal for engineering students, within the context of essentially traditional concepts of chemical science represents a meaningful compromise. Students appear to have a greater buy-in for the course because the benefits towards their own goals for their studies are made explicit. At the same time, it is possible to convince students that problem-solving in any domain requires fundamental knowledge of the content – in this case chemistry.

The trajectory by which this problem-solving emphasis has emerged for this course has been presented here. To be sure, a significant amount of research remains to establish that the goals of transferring problem-solving skills are achieved with this approach. Such research is expected to be initiated in the future.

Acknowledgments

We acknowledge several different grants from the National Science Foundation over the course of the trajectory described here that allowed this long-term look for this project to be possible. Specific grants were: DUE-9752280, DUE-0618600, DUE-0817409. Assistance from R. Stevens in the assignment of quadrant scores for the IMMEX study is gratefully acknowledged.

References

1. Holme, T. *J. Chem. Educ.* **2001**, *78*, 1578–1581.
2. Bressoud, D. M. In *Calculus: The Dynamics of Change: MAA Notes #39*; Roberts, A. W., Ed.; Mathematical Association of America: Washington, DC, 1996.
3. Keynes, H. B.; Olson, A. M. *Int. J. Math. Educ. Sci. Technol.* **2010**, *31*, 71–82.
4. Fox, E.; Riconscente, M. *Educ. Psychol. Rev.* **2008**, *20*, 373–389.
5. Sandi-Urena, S.; Cooper, M. M.; Stevens, R. H. *Int. J. Sci. Educ.* **2011**, *33*, 323–340.
6. Miller, T. M.; Geraci, L. *Metacognit. Learn.* **2011**, *6*, 303–314.
7. Nurrenbern, S. C.; Pickering, M. *J. Chem. Educ.* **1987**, *64*, 508–510.
8. Nakhleh, M. B.; Lowrey, K. A.; Mitchell, R. C. *J. Chem. Educ.* **1996**, *73*, 758–762.
9. Sanger, M. J.; Greenbowe, T. J *J Res. Sci. Teach.* **1997**, *34*, 377–398.

10. McDermott, L. C.; Shaffer, P. S. *Am. J. Phys.* **1994**, *62*, 46.
11. Pfister, H.; Laws, P. *Phys. Teach.* **1995**, *33*, 214.
12. Tribe, J. *Educ. Rev.* **1996**, *48*, 13–27.
13. Chang, L.; Dziuban, C. D.; Olson, A. H. *Appl. Meas. Educ.* **1996**, *9*, 161–173.
14. Jaefer, R. M.; Tittle, C. K. *Miniumum Competency Achievement Testing: Motives, Models, Measures and Consequences*; McCutchan Publishing: Berkeley, CA, 1980.
15. Holme, T. *J. Chem. Educ.* **2003**, *80*, 594–598.
16. Brown, L.; Holme, T. *Chemistry for Engineering Students*; Brooks/Cole, Cengage Publishing: Belmont, CA, 2005.
17. Schultz, M. J. *Chemistry for Engineers: An Applied Approach*; Brooks/Cole, Cengage Publishing: Belmont, CA, 2006.
18. Glanville, J. O. *General Chemistry for Engineers*, 2nd preliminary ed.; Prentice-Hall: Upper Saddle River, NJ, 2002.
19. Bodner, G. M. *Univ. Chem. Educ.* **2003**, *7*, 37–45.
20. Stevens, R. H.; Soller, A.; Cooper, M.; Sprang, M. In *7th International Conference Proceedings*; Lester, J. C., Vicari, R. M., Paraguaca, F., Eds.; Springer-Verlag: Berlin, 2004; pp 580−591.
21. Cooper, M. M.; Cox, C. T.; Nammous, M.; Case, E.; Stevens, R. *J. Chem. Educ* **2008**, *85*, 866–870.
22. Stevens, R.; Palacio-Cayetano, J. *Cell Biol. Educ.* **2003**, *2*, 162–179.
23. Stevens, R.; Johnson, D. F.; Soller, A. *Cell Biol. Educ* **2005**, *4*, 42–57(2005).
24. Stevens, R. H.; Thadani, V. *Technol., Instr., Cognit., Learn* **2007**, *5*, 325–337.
25. Bowen, C. *J. Chem. Educ.* **1994**, *71*, 184–190.
26. Calimsiz, S. *Problem-Solving Cannot Be Taught: We Can Teach Gap Reducing Techniques*; Purdue University: West Lafayette, IN, 2003.

Chapter 6

Developing a Content Map and Alignment Process for the Undergraduate Curriculum in Chemistry

April L. Zenisky[1] and Kristen L. Murphy[*,2]

[1]Center for Educational Assessment, Department of Educational Research, Policy, and Administration, University of Massachusetts-Amherst, Amherst, Massachusetts 01003
[2]Department of Chemistry and Biochemistry, University of Wisconsin-Milwaukee, Milwaukee, Wisconsin 53201
[*]E-mail: kmurphy@uwm.edu

Faculty and departments are regularly requested to propose methods to measure student performance as part of programmatic assessment and accreditation. In addition to classroom assessment efforts, these programmatic assessment efforts can be approached by comparing students to national samples of students using a norm-referenced exam. However, these results give limited insight into what students know about the content and, more importantly, how this knowledge changes over an undergraduate program. Approaching the reporting of student performance from the perspective of what students know requires the exam items are designed or aligned along a framework of content and complexity. For college-level chemistry, this required the development of the content portion of this framework. The process for this development deliberately involved wide-spread contribution from practicing experts (faculty and instructors) in the community. The iterative process provided a robust, detailed structure and content map that allows faculty and departments to gauge student learning by criteria in addition to comparison to a national sample. The process for the development of this map, the process of aligning chemistry exam items, and the implications for the use of the map is described.

© 2013 American Chemical Society

Introduction

Classroom instruction is complemented by classroom assessment. Through better measurements of student content knowledge, instructors have a better ability to judge the degree to which individual students or classes have learned course content. Classroom assessment efforts can also complement programmatic assessment efforts where individual students or cohorts of students can be regularly evaluated for growth in their content knowledge. However, much of this assessment and evaluation is (typically) still isolated within individual institutions. In order to gauge the knowledge of students in a national sample, standardized testing can be used. On the college level for chemistry, the American Chemical Society, Examinations Institute (ACS-EI) has been providing high-quality exams for over 75 years. For much of this time, these tests have also been developed as *norm-referenced* assessments, where performance of individuals and groups is compared to demographically-representative national samples (the 'norm group'). This approach to reporting results provides performance data to the community that supports intended users of the scores in gauging the performance of students at any one institution relative to a national sample. There is value in this ability to support comparative interpretations for students nationally, but this measure does not give insight about what individual students or classes actually know overall or within a particular content area. When student test performance is compared to an absolute standard of knowledge and skills (rather than against the performance of other test-takers), the test fulfills a *criterion-referenced* purpose, where the criterion is defined as mastery of 'standards', 'benchmarks', 'indicators', or 'skills', as defined by the framework underlying the assessment.

In order to gauge how well students perform within a content area according to an absolute standard (the criterion-referenced approach), exam items must be organized by the specific content that is tested. Traditionally, this has been done by using National Standards (*1*) and aligning test items to these standards (*2*). Additionally, the standards can provide a framework for specifying the content of an assessment during test development (*2*). On the college level in STEM fields, no national standards exist for science generally or chemistry specifically. This paper describes the process undertaken over a three-year period to develop a framework of the undergraduate curriculum in chemistry and the use of this framework to map existing test items by content. Figure 1 shows both the overall structure for this map (by levels) and the generic process by which the general chemistry content map was constructed (with more detail on the particular focus groups and other subdisciplines shown in Table I). This process is in progress, and will ultimately result in the alignment of chemistry exam items from ACS-EI exams to the full framework in chemistry disciplines; this will also support criterion-referenced reporting of individual students or classes to the intended users of ACS-EI exams (typically, faculty and departments, but also potentially institutions as part of periodic programmatic evaluation activities). Additionally, the content maps (*3*) and the process of aligning exam items (described here) are available to the community for aligning other exams items. It is also expected that these alignments may be used for both classroom and programmatic assessment efforts where greater value may be placed on what students know instead of how students rank nationally. Lastly, the process by which ACS-EI develops exams is decided by the committees that write these exams (*4*). It is not the intent of the ACS-EI to impose these content maps as a framework or architecture for test design. The choice of exam committees on how to use the content map will reside with the individual committees. Alignment of the exam items written by any exam committee is expected to follow the test design.

Content Map Structure:

Level 1	Level 2	Level 3	Level 4
Anchoring Concepts or "Big Ideas"	Enduring Understandings	Sub-disciplinary Articulations	Content Details
sub-discipline independent		sub-discipline specific	

Process for Development (General Chemistry):

Level 1	Level 2	Level 3	Level 4
		Focus Group I: *Synthesis* Focus Group II: *Testing/Refinement*	Focus Group III: *Selection and Analysis*
Initial synthesis occurred prior to the development of general chemistry content map. *Further vetting continuously occurs (particularly during the first focus groups focusing on Level 3 synthesis)*		Following Focus Groups: *Testing/refinement through trial and actual alignment processes*	

Figure 1. Content map structure and the generic process for development (shown here for general chemisty).

Background

The process of building a test can perhaps best be described as one marked by varying levels of formality, depending on the level of the stakes associated with the test. For classroom assessment purposes, an instructor can develop a test to assess student knowledge of interim or summative course content through a considerably more straightforward and simple evaluative activity without much (if any) involvement from others. In other testing contexts, such as admissions, credentialing, graduation, and accountability testing, the process is differently conceived of and managed as a sequence of steps all based on decades of research and practice in the field of psychometrics. These all are designed to provide evidence to support the proposed inferences to be made on the basis of test scores. For example, in the credentialing of doctors, the process of test development is handled by psychometricians who draw on the expertise of doctors in formulating all aspects of the test, including determining what the content of the test should be, how questions should be crafted, and what standards are appropriate for passing. In this way, validity evidence is obtained by both procedural means as well as subject-matter expert participation. In reflecting on the ACS-EI's processes of test development described in the present paper, it should be noted that the model that has been used by the ACS-EI since its inception has been both cooperative and focused on grass-roots participation, and the present work carries on that legacy of inclusion (*4*, *5*).

Developing tests to measure academic knowledge and skills in high-stakes settings such as K-12 and higher-education assessment requires that agencies and test developers define the domain to be assessed. This can and should be considered to be a critically essential process that is composed of general steps, but one that is decidedly not one-size-fits-all in terms of the mechanisms or strategies used to determine the key elements that form the basis of all subsequent test development. What is important is that whatever process is used be formulated with the singular intent of providing evidence of content validity to support the proposed uses of the assessment and appropriate uses of test scores (*6*). In this case, test content involves test elements such as how questions are formulated and worded to measure certain knowledge, skills, and abilities (*6*).

In the broadest sense, the steps involved in construct definition (which occur prior to item development) require the formulation of an overall plan, a process of content definition, and the development of test specifications (*6*). Such a systematic approach to assessment design is intended to promote reliability, validity, and indeed quality throughout the test-development process.

In terms of the overall plan, the types of considerations that are discussed here address providing high-level guidance for all subsequent test development work, including construct definitions, articulation of the purpose(s) (and, derived from that, intended use(s)), and format determination (*6*). As described in the *Standards for Educational and Psychological Assessment* (*7*), the testing purpose and the intended use(s) of the assessment are the fundamental basis for judgment about the sufficiency of the assessment to accomplish those goals. A clear statement of the purpose of testing is essential to provide continuity among test specifications, domain boundaries, and individual test items, all of which support validity. In other words, the process provides the test developers the ability to focus the assessment items on the relevant content to test. The process itself reveals these judgments through the product (the assessment) that is developed.

With the fundamental aspects of the domain in hand, the next step is to delineate the domain to be assessed, in terms of content, skills, processes, and features. This can be accomplished through a wide variety of strategies. Credentialing examinations typically rely on practice analysis (also referenced as job analysis), but indeed the options are more open-ended in other types of assessment (*6*). In classrooms, instructor judgment about what to test is ideally reasonable and sufficient. Instructors draw upon their expertise of the domain and of their students to make these judgments. These judgments can then translate into the creation of exam items or the use of exam items that were created by other experts. The process of writing, editing, testing and selecting exam items can be developed through professional development activities such as serving on exam committees (*4*).

Focusing on high-stakes achievement testing contexts, the process matters as much as the outcome, where defensibility is a paramount concern. Typically, content definition and specification require both a scientific aspect that involves standardized procedures for collecting and synthesizing sources and an "art" portion in which the proposed content basis of a test is evaluated (*7*, *8*). Systematically, Webb (*2*) notes that a test's specifications are based on 1) the content topics, 2) the complexity of test items, 3) the range of content to be covered, and 4) the degree of emphasis to be given to various content expectations. Lastly, while it must be noted that human judgment occupies a significant role in how these elements are determined, such judgment can be harnessed to promote defensibility of process. Strategies for this include instituting practices for promoting the dependability of subject-matter experts, documenting the qualifications of the participants, and ensuring the adequacy of the methods used.

One particular approach to domain definition is the creation of a content map. In this approach, subject matter experts collaborate with test developers to identify the domain to be assessed framed in a hierarchical structure, as is appropriate for the domain. In the abstract, consider a mathematics assessment. The high-level topics might include Numbers, Algebra, Geometry and Measurement, and Statistics and Probability. Within each of those topic areas, the experts would identify subdomains and then specific standards associated with each subdomain. In this way, the map provides a visual representation of the domain.

Next, the alignment study process is a mechanism by which domain definition is checked against the contents of a test. As with domain definition and specification, there is no single strategy for alignment that must be followed; rather, the choice of approach can be weighed against the test purpose. Some widely-used ways of thinking about alignment include (*2*):

- Categorical concurrence: The major topics on a test should match up with the major topics in content standards, and the standards therein should be measured at a sufficient level of depth to permit reporting at the topic level.
- Depth of knowledge: The complexity of standards can be compared to complexity of items to check for stability.
- Range of knowledge: This dimension is focused on the breadth of the items included on an assessment and how that breadth relates to the possible range of standards that could be represented on an assessment, to establish the range of standards representation on an assessment (and hence domain representation).

There are numerous examples of how this process is carried out in practice for academic content areas, in both higher education and K-12 settings. One example involves the Major Field Tests, developed by Educational Testing Service (*9*). These assessments are used by higher education institutions to assess undergraduates' skills, knowledge, and understanding in a number of undergraduate fields of study. The domain is defined using specifications developed from national curriculum surveys, and reviewed by subject-matter experts. A similar example for the process is the development of the College-Level Examination Program (CLEP) by the College Board. Curriculum surveys are used to gather information about the primary content and skill areas covered in courses (including proportional time allotment), topics taught and the emphasis given, expectations, and textbooks and resources used in teaching (*10*). Curriculum survey materials are then used by test development committees to formulate exam content specifications, among other test development tasks (e.g., item development and selection).

Methods

The development of the content map for the undergraduate chemistry curriculum specific to sub-disciplines took place over a three-year period. In order to work with a group of instructors in the field of chemistry, the work was conducted at regional, divisional, and national meetings or conferences. The meetings, locations, date and focus group(s) are listed in Table I.

Each session was three hours in either the morning or afternoon of the meeting or conference. Multiple sessions rarely ran concurrently, allowing the participants to be present at more than one session and the facilitators to be present at all sessions. Each session began with an introduction of how measurements made by standardized exams provide information to instructors or programs about student proficiency (either about an individual or a group). The discussion then focused on ACS Exams and the value of norm-referenced assessments. A discussion of criterion-referenced exams would then be introduced as well as the value associated with this information for both classroom and programmatic assessment. The discussion naturally led to the need for criteria or standards in order to align ACS Exam items for criterion referencing. The development of the content portion of this map was then introduced as the objective of the focus group. A description of the structure of the map (shown in Table II and Figure 1) was presented with the specific task for the group then described.

Table I. Timeline of the development of the sub-discipline specific statements for the content map

Date	*Meeting or Conference*	*Focus Group*
July, 2008	Biennial Conference on Chemical Education (Bloomington, IN)	Level 3 synthesis (General Chemistry)
		Testing and refinement (General Chemistry map)
March, 2009	ACS National Meeting (Salt Lake City, UT)	Level 4 selection and synthesis (General Chemistry)
		Testing and refinement (General Chemistry map)
		Level 3 synthesis (Organic Chemistry)

Continued on next page.

Table I. (Continued). Timeline of the development of the sub-discipline specific statements for the content map

Date	*Meeting or Conference*	*Focus Group*
August, 2009	ACS National Meeting (Washington, DC)	Testing and refinement General Chemistry map
		Level 3 synthesis and refinement (Organic Chemistry)
October, 2009	ACS Regional Meeting (Hartford, CT)	Testing and refinement (General Chemistry map)
March, 2010	ACS National Meeting (San Francisco, CA)	Testing and refinement (General Chemistry map, now using first and second term General Chemistry exam items)
		Testing and refinement (Organic Chemistry map)
August, 2010	ACS National Meeting (Boston, MA)	Testing and refinement (General Chemistry map, now using first and second term General Chemistry exam items)
December, 2010	ACS Regional Meeting (New Orleans, LA)	Testing and refinement (Organic Chemistry map)

Table II. Structure of the four levels of the content map

Level 1	Anchoring Concept	Sub-discipline independent
Level 2	Enduring Understanding	Sub-discipline independent
Level 3	Sub-disciplinary Articulation	Sub-discipline dependent
Level 4	Content Details	Sub-discipline dependent

Participants were sought through email prior to the meeting or conference. The email was sent to current and prior ACS-EI volunteers as well as attendees who selected "academia" in their registration. Participants were invited to attend sessions appropriate to their expertise and current teaching experience. A typical focus group had six participants with a maximum of fifteen. The participants were provided with written copies of the current content map with space for their contribution. At the conclusion of each session, all materials were collected for analysis. All data was entered into the content map based on the consensus of the participants and with minor revisions or synthesis done internally by ACS-EI personnel.

Tasks and Results

The structure of the map (given in Table II and shown in Figure 1) began with eight broad statements (Big Ideas or Anchoring Concepts) at the first level that encompass the entire undergraduate curriculum in a typical chemistry program. These were evaluated early in the process and two additional Anchoring Concepts (experimental and visualization) were proposed and included in the map. These were then parsed into smaller statements (Enduring Understandings) at the second level that were still sub-discipline independent or applicable to all levels within the undergraduate curriculum of chemistry. The third level was the first point at which the sub-discipline specific application of the map began with the Level 3 statements (Sub-disciplinary Articulations) giving a finer detail to the Level 2 statements. The final level statements (Content Details) were the finest level of detail, and it is to these statements that individual test items would be aligned. The map for general chemistry has been published as well as the results from initial alignment studies using the general chemistry map (*3*, *11*).

The process of developing the domain map for the undergraduate chemistry curriculum began with the outline for the map. This was then utilized to design the first iteration of the Level 1 and 2 statements that are sub-discipline independent. The majority of the work on the domain map was focused on working on the Level 3 and 4 statements where the input from the instructors in the field teaching these courses was most important. These iterations of the map were tested through multiple "trial mapping" sessions (called "refinement" in Table I) where items from exams were placed within the domain map based either on Level 3 or 4 statements.

The initial focus of the development of the Level 3 statements was on the sub-discipline of general chemistry. At the first focus group, in small groups (of two or three), participants were provided with the Level 1 and 2 statements and were invited to comment on these (both for content and context) as well as determine if these were applicable to general chemistry (by simply checking a "vital to general chemistry" box). The participants then wrote the corresponding Level 3 statement(s). These statements were not restricted to number or length; there could be any number of Level 3 statements corresponding to a single Level 2 statement.

The participants were very engaged and enthusiastic with the project. They not only understood the immediate task but also saw the extension into their teaching. For some this was a "confirmation" of their somewhat unique approaches (for instance, one participant explained how the anchoring concepts and enduring understandings were appropriately similar to their approach in the classroom). Others initially tied the map to the current order of their course or even the textbook they used. Through the initial discussion, these participants were drawn by others to abandon the order of the textbook and approach the task from the perspective of generically identifying how general chemistry fit within this framework (which they were eager to do). Overall, the group work on this process was key as well as the composition of the groups themselves. Attempting to have a balance between traditional versus progressive teaching approaches was important.

It appeared easier to work one Big Idea at a time: consider the Big Idea, consider/revise/add the Enduring Understandings, and finally add the Sub-disciplinary Articulations. The discussions about the Level 2 statements (Enduring Understandings) included more comments about editing or revising the statements. There seemed to be some difficulty in not adding too much detail or specifying the Level 2 statements for general chemistry. Additionally, once the discussion turned to the Level 3 articulations, again the participants struggled with adding too much detail or restating the Level 2 statement. There seemed to be a desire to make the Level 2 statements into shorter more succinct statements and add additional Level 2 statements to deal separately with the ideas previously tied into one. In a few cases this was appropriate and the group agreed on the new set of statements.

The first draft of the domain map for general chemistry was assembled through Level 3 statements. The majority of the statements generated by the groups at the first focus group were in good agreement and little discussion or editing by ACS-EI personnel was required for consensus on inclusion. These Level 3 statements were then trial-tested in a "trial-mapping" session at the second focus group where general chemistry items from a inactive ACS General Chemistry Exam were assigned to a position in the content map. Through this process, it was evident where obvious gaps were by which exam items could not be assigned. Additionally, it was evident that some Level 3 statements were too broad or too narrow and more editing was necessary before the Level 4 statements could be generated. The editing was done by ACS-EI personnel prior to the next focus group.

The next phase of the development of the general chemistry map focused on adding the Level 4 statements. Because it was expected that these would be specific to the level of a single standard test item, the Level 4 statement outline began not with the map or a typical exam but with the sub-topic headings in a typical general chemistry textbook. The number of sub-topic headings in any textbook encompasses much more specific content areas than would be expected to be taught or assessed. Therefore, the headings were assigned into five broad categories and color-coded as shown in Table III. These assignments were initially made by personnel at ACS-EI and included how often these specific content areas are tested or how important these are to test. These ratings would ultimately serve as the basis for inclusion of the specific content areas into the Level 4 statements.

The color-coding rather than the levels were given to different group of participants at the third focus group, and they were instructed to react to the initially coding of these sub-topic headings based on their testing practices (rather than using a standard ACS General Chemistry Exam). The participants worked on this alone but the session concluded with a discussion on both the assignment of the sub-topic headings and the reflection of the individual testing practices of the participants. The participants were not provided with the domain map during this process but were informed of the domain map and the inclusion of the Level 4 statements into the map. The color-coding assigned by each participant was translated into a level assignment (consistent with Table III) and these were combined with the original assignment for a ranking based on the median value given. Only the highest three values were considered for inclusion into the content map and, in some cases, the sub-topic headings were too specific and were combined into one Level 4 statement.

The initial draft of the Level 4 statements was again trial-tested through a "trial-mapping" session. Participants worked in small groups and placed ACS General Chemistry Exam items using all four levels in the content map. Changes to the Level 3 and 4 statements were proposed as well as a discussion about the utility of the Level 4 statements. It was proposed that the items could be mapped without the Level 4 statements. The changes proposed to the content map were included in the next draft. Three more "trial-mapping" sessions took place with different ACS General Chemistry Exams. During these sessions, minor changes were proposed for the map as well as the need for the Level 3 statements but not the Level 4 statements. The most recent version of this content map has been published (*3*).

Table III. Five levels of assignment of sub-topic headings from a typical general chemistry textbook

Level	*Color-coding*	*Importance of testing*	*Frequency of testing*
1	Green	Important to test	Always or almost always included
2	Blue	Unimportant to test	Always or almost always included
3	Yellow	Important to test	Sometimes test
4	Pink	Important to test	Rarely or never test
5	White	Unimportant to test	Not tested

Concurrent to the development of the general chemistry content map, work began on the organic chemistry content map. Through three focus groups, participants generated and edited Level 3 statements specific to organic chemistry. Again these drafts were "trial mapped" with ACS Organic Chemistry items. Through these mapping sessions, it was found that a slight change to one Level 2 statement was necessary that did not affect the general chemistry content map. Work is ongoing in developing the specific components of the content map for the other subdisciplines (biochemistry, physical chemistry, analytical chemistry and inorganic chemistry) with varying degrees of completion on these maps at this time. Because it was never intended that the content portion of the map will be static, as the work in these areas is completed, it is expected that there will be minor modifications to the Level 2 statements. These modifications will be retrofitted to all completed content maps including general chemistry.

The overall process to develop these maps results in content maps that can be used to align chemistry exam items to a framework of content and complexity. This framework can be used to evaluate student performance based on content rather than comparatively to other students who took the test. For example, by mapping a group of items in chemistry to the first Anchoring Concept and first Enduring

Understanding, the performance by a student or a group of students on these items can be used to evaluate performance in the larger context of what students know about this big idea in general chemistry and how this extends beyond general chemistry into the other sub-disciplines of chemistry (Figure 2).

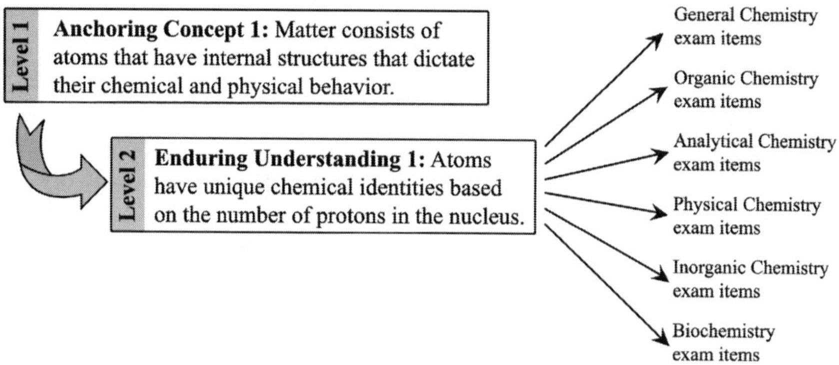

Figure 2. Alignment of chemistry exam items within the content portion of the map.

The ACS-EI has begun the process of mapping items from general chemistry (*11*) and organic chemistry. These alignments are conducted by chemistry instructors and the items are aligned by both content and complexity. The content alignment is conducted through the use of the content map. The assignment of complexity is introduced using a published rubric (*12*) and, within this context, the raters assign the items as easy, medium or hard. The alignment process can also be used to examine the content coverage on the exam by looking for a lack of items within an Anchoring Concept. For example, within general chemistry I, it would be expected that few items would be placed within the Anchoring Concept of "kinetics" as this is typically covered in general chemistry II.

These alignments have been integrated into the reporting of electronic testing using ACS exams. For example, standard reports include both raw total scores and percentile rank of students. New reports will now include sub-scores based on Anchoring Concepts. Furthermore, instructors will also have the option of examining finer detail into Enduring Understandings, Sub-disciplinary Articulations or even Content Details. Additionally, instructors will have the option of viewing individual students' sub-scores in all Anchoring Concepts.

The content maps and the process for aligning exam items have been and will be available to the community (*3*, *11*). Programs using ACS-EI exams will have the option of considering student performance using norm-referenced or criterion-referenced reporting. Additionally, instructors can use the framework to align other non-ACS-EI items for examining student performance by content area. Beyond this, instructors or programs can use or adapt the framework to meet their assessment needs. For example, extending this practice into multiple exams over the undergraduate program, one can examine again for the content knowledge within a specific Anchoring Concept and how this changes over time (and through instruction in various courses) or coverage of content over the course of the undergraduate program. This provides an opportunity to use the content map for both classroom and programmatic assessment efforts. This also allows for longitudinal studies of specific cohorts of students.

Discussion

The process for the development of the content map for the undergraduate chemistry curriculum was unique. This was due to the absence of an administrative role or top-down approach for creating or imposing a content map. Similarly, no externally imposed standards exist for the undergraduate chemistry curriculum to drive the process. However, the necessity for a model for higher-education accountability as well as an opportunity to investigate what students know provided the motivation to develop the map. Accordingly, the process was developed for the instructors in the field where the participants were immediately active in the process and their input was integrated into the results. Often the process invited vigorous discussions between instructors with different opinions or different teaching approaches. However, the process was not driven by a few and the many contributions from many instructors were integrated into the final version of the content map. Additionally, the utility of the content map was often immediately recognized by the participants. They appreciated both the importance of measuring student proficiency by criterion-referencing and the use of the content map as a universal tool for programmatic assessment. Beyond this, once items either from standardized testing or instructor-created exams are aligned to the framework, the ability to better understand student performance and act on this information is possible. The use of exams as diagnostic instruments to better prepare students throughout their undergraduate program is enhanced through efforts such as these.

Finally, the process described here is not the first or last attempt to capture the content of the undergraduate curriculum in chemistry. However, the process utilized by the ACS-EI was deliberately designed to be inclusive and participatory, and this approach is guided by and reflective of best practices in assessment more generally. When subject-matter experts are involved in all aspects of test development, this speaks to the validity evidence of the process, and ultimately the outcome (the assessment). To this end, in the present context, the efforts to draw on expertise in the field were intentional.

Also, it should be noted that the map was iteratively developed and revised by many faculty and instructors. The map was vetted through multiple processes including creation, editing, and trial alignments, and this was done by considering the lowest level of detail from a course, not a particular exam. From a methodological perspective, much of this process was carried out using multiple focus groups, meaning participants were asked to participate in a collaborative and respectful process to craft the maps. Every effort by the ACS-EI was made to include all experts who wanted to be part of the process and not anchor the process to any particular exam, course, textbook, institution or teaching methodology. Because of this process, it is expected that the community will find great value in this map, as the community was an integral part in its development. The ACS-EI would not be able to produce the exams that it does nor share these innovative approaches to measuring student learning without this community.

Acknowledgments

We gratefully thank Thomas Holme for his innumerable and invaluable contributions to this process, the ACS-EI and to the chemistry community. We also gratefully thank all of the participants in the focus groups without whom this process would not have been possible. Lastly, we thank the chemistry community for its active involvement in valuing the measurement of what students know and how to constantly improve the ability to measure this.

References

1. N. R. C. National Committee on Science Education Standards and Assessment *National Science Education Standards*; National Academies Press: Washington, DC, 1996.
2. Webb, N. In *Handbook of Test Development*; Downing, S. M., Haladyna, T. M., Eds.; Lawrence Erlbaum Associates, Inc.: Mahwah, NJ, 2006; pp 155−180.
3. Holme, T. A.; Murphy, K. L. *J. Chem. Educ.* **2012**, *89*, 721–723.
4. Holme, T. A. *J. Chem. Educ.* **2003**, *80*, 594–598.
5. Haertel, E. H. *Educ. Meas.: Issues Pract.* **2002**, *21*, 16–22.
6. Downing, S. J. In *Handbook of Test Development*; Downing, S. M., Haladyna, T. M., Eds.; Lawrence Erlbaum Associates, Inc.: Mahwah, NJ, 2006; pp 3−25.
7. American Educational Research Association (AERA), American Psychological Association (APA), and National Council on Measurement in Education (NCME) *Standards for Educational and Psychological Assessment*; AERA: Washington, DC, 1999.
8. Knupp, T., and Harris, D. J. In *Handbook on Measurement, Assessment, and Evaluation in Higher Education*; Routledge: New York, 2012, pp 236−256.
9. Educational Testing Service. Major Field Tests: Content. Educational Testing Service: Princeton, NJ, 2013. http://www.ets.org/mft/about/content/.
10. College Board. Test Development, CLEP. College Board: Newtown, PA, 2013. http://clep.collegeboard.org/test-development.
11. Murphy, K. L; Holme, T. A.; Zenisky, A. L.; Caruthers, H.; Knaus, K. J. *J. Chem. Educ.* **2012**, *89*, 715–720.
12. Knaus, K. J.; Murphy, K. L.; Blecking, A.; Holme, T. A *J. Chem. Educ.* **2011**, *88*, 554–560.

Trajectories of Reform of Teaching Tools and Methods

Chapter 7

PLTL: Tracking the Trajectory from Face-to-Face to Online Environments

Pratibha Varma-Nelson*[,1,2] and Julianna Banks[2]

[1]Department of Chemistry and Chemical Biology, Indiana University-Purdue University, 755 W. Michigan Street, Indianapolis, Indiana 46202
[2]Center for Teaching and Learning, Indiana University-Purdue University, 755 W. Michigan Street, Indianapolis, Indiana 46202
***E-mail: pvn@iupui.edu**

Over the past three years, an interdisciplinary team of investigators, led by Varma-Nelson, has worked to adapt the Peer-Led Team Learning (PLTL) instructional model to a cyber-environment (aka cPLTL). PLTL is a pedagogy that preserves the lecture and replaces the course recitation with a weekly two-hour workshop in which six to eight students work collaboratively to solve challenging problems under the guidance of a peer leader. cPLTL is the "cyber" evolution of PLTL to an online format. The team's work represents a new direction for educational research and expands the knowledgebase on teaching science, technology, engineering and mathematics (STEM) concepts, while using technology as an educational tool. With funding from the National Science Foundation (NSF) and the Next Generation Learning Challenges (NGLC) initiative, the team is examining cPLTL's impact on student performance. Analysis of course grades and standardized exam scores has shown cPLTL's positive impact on educational outcomes. This chapter traces the evolution of a pedagogy developed for the face-to-face classroom environment to an online platform. Specifically, it outlines the rationale that led to the development of cPLTL; describes how technology was integrated into the PLTL model; summarizes its effectiveness, outcomes, and lessons learned; and speculates on the future use of cPLTL.

© 2013 American Chemical Society

Introduction

Peer-Led Team Learning (PLTL) is a nationally recognized model for teaching and learning in science, technology, engineering and mathematics (STEM) disciplines. Grounded in social constructivism (*1*) and social interdependence theory (*2*), the model preserves the lecture and replaces the recitation with a weekly two-hour workshop session. During the interactive workshops, six to eight students work as a team to solve carefully constructed problems under the guidance of a trained peer leader. Within this model, peer leaders are undergraduates who have successfully completed the course and are trained to facilitate discussions and engage students (their peers) in problem solving activities in which they discuss, debate and defend their decisions and reasoning in complex problem-solving (*3*). The workshops make it possible for students to engage with course material and with each other in collaborative discussion and constructive debate as they develop effective problem-solving strategies. The workshops also facilitate the development of important workplace skills as students learn to obtain answers by recognizing, evaluating, and choosing the best solutions. This process requires that they collaborate, analyze, explain, negotiate, modify, listen, learn solutions, and deliver tangible work products as a team, all of which are transferrable (*4*), marketable skills.

Under the guidance of David Gosser and his colleagues, the model was originally introduced as *Workshop Chemistry* in the City College of New York's first-year general chemistry course. In 1995, it became one of five systemic change initiatives (NSF-DUE-9455920) (see Susan Hixson's chapter in this book; Chapter 2). This grant was followed by a national dissemination grant in 1999 (NSF-DUE-9972457). A supplement to this grant assisted in propagating the model to community colleges. A second national dissemination grant was awarded in 2003 (NSF-DUE-0231349) to facilitate the formation of regional PLTL centers. A centerpiece of both dissemination grants was the use of small grants called Workshop Project Associate (WPA) grants. 92 of these mini-grants were awarded in a variety of STEM disciplines.

Following the dissemination of PLTL, a compelling body of evidence was produced demonstrating that student performance can be significantly increased (with an average gain of 15% in students earning A, B, or C grades, regardless of gender or race) with "a relatively straight-forward modification of teaching style, that incorporates peer-led workshops to complement either a reduced lecture or recitation or as an added component" (*5*).

As *Workshop Chemistry* expanded, it was later renamed *PLTL* by the consortium formed through NSF funding. While originally developed for chemistry instruction and learning (*6–9*), PLTL is now used in biology (*9–11*), engineering (*12*), mathematics (*9*, *13*), physics (*9*), psychology (*9*, *14*), computer science (*15*, *16*), nursing (*17*), and business (*18*), as well as other fields. An introduction to the model, its history, an overview of the PLTL project, and its evolution from *Workshop Chemistry* to *PLTL* is documented in the 2008 James Flack Norris award address (*19*).

Implementation guidelines for the model are detailed in the *PLTL Guidebook* which describes leader training, materials development, faculty role, and the theoretical underpinnings of the model (*20*). The evolution of PLTL into a national STEM educational innovation and a detailed description of the critical components for successful implementation as well as limits to adaptability of the model are discussed in a monograph by Gafney and Varma-Nelson (*21*). This book also provides a detailed description of the evaluation, dissemination, and factors that lead to institutionalization of PLTL. The evaluation methods outlined in this latter publication are transferrable to other STEM pedagogies.

The impact of PLTL workshops on students, leaders, faculty and institutions has been assessed, evaluated, and documented in a variety of settings for nearly two decades (*22–28*). The results are consistent and consequential for undergraduate education. Students and leaders participating in the PLTL workshop model are more successful than non-workshop students. But is this where the PLTL story would end? Has PLTL's potential to enhance how and what students learn been fully explored? This chapter details the catalysts that led to the development of cPLTL; how technology is integrated into the PLTL model; and its effectiveness, outcomes, and lessons learned.

Rationale for Developing cPLTL

The landscape of higher education has undergone significant change over the last three decades. Changes in the economy, education policy, and social structures have resulted in larger and more diverse classrooms (*29*). The changes in student demographics and advances in technology have prompted institutions to: 1) rethink how instruction is delivered, and 2) to foster and develop innovative practices that further enhance student learning and development. These initiatives have challenged us to reconsider how we think about PLTL instruction, its delivery, current limitations, and the potential impact of integrating technology.

Enrollment in online courses in the United States has been growing substantially faster than higher education's overall enrollment (*30*). During the Fall 2007 term, nearly 3.9 million students took at least one online course. This represented 21.9% of the total enrollment that year which was up from 9.9% in Fall 2002 (*30*). So, the idea of moving PLTL online was timely and held the potential to reach populations excluded from PLTL participation. Public institutions and community colleges had the highest rate of increase in online courses including in the science disciplines (*30*). Focusing cPLTL testing and adoption on public universities and community colleges was ideal because they were uniquely positioned to extend access and the positive outcomes of cPLTL to a broader student population.

Higher education has been among the hardest hit in the recent economic downturn with a seven percent decrease in state and local support for higher education in 2012 (*31*). During the dissemination phase of the national PLTL project, not all who desired to adopt PLTL in their courses could do so for a variety of reasons including: lack of adequate classroom space for groups, difficulty with scheduling an additional two hours during the day, and the availability of leaders, especially at two-year institutions. Because of the economic downturn and institutions' increased interest in managing costs and resources, cPLTL was seen as a viable low-cost technology solution.

Today, a majority of all students attending higher education commute to school (*32*, *33*). Commuter campuses hold a unique position in regards to student interactions, as classrooms are typically the only regular settings that their students have for interaction with faculty and peers (*33*). If the impact of college is determined by the type and amount of interaction with major socializing agents (faculty and peers) on campus and peers are the single strongest source of influence on cognitive, affective, psychological, and behavioral development (*34*, *35*), then curricular solutions should intentionally create communities of learning that maximize and encourage continuous student interactions in and outside of the classroom. This presents challenges for commuter institutions and those with mostly online programs. PLTL allows for high quality educational activities that require students to dedicate substantial time and effort toward educationally purposeful tasks (*36*). The workshops place students in settings and situations that require regular interaction with peers on substantive issues, over extended periods of time; and now, with cPLTL, in spaces beyond the classroom.

Creation of cPLTL holds the potential to increase participation and success of all students in STEM fields by providing active learning and leadership opportunities to a more diverse group of students in a flexible time frame. Essentially, it provides opportunities for those who perform well in the course, but are unable to serve as PLTL leaders in traditional settings, to serve as peer leaders in the online environment. The availability of cPLTL permits the PLTL pedagogy to be more widely adopted and implemented, especially in urban commuter universities because it:

- eliminates the need to locate appropriate classrooms for each group, as workshops can be scheduled in any time slot during the day or night;
- allows students to develop additional workplace skills—information and computer literacy and digital collaboration skills; and
- provides flexible scheduling, for working students and those with families who are unable to enroll in workshops or participate as leaders due to conflicts with work and family schedules.

Considering the trends in student enrollment and recent investments in developing online courses, the next logical step was to take PLTL to an online platform.

cPLTL Development

PLTL's documented success in improving student achievement and retention led Indiana University Purdue University Indianapolis (IUPUI) to implement it in general chemistry in 1998 (*19*, *21*). By 2008, when Varma-Nelson was appointed Executive Director of the Center for Teaching and Learning (CTL), IUPUI's PLTL program had achieved sustained success with the number of students receiving D and F grades in fall semesters decreasing from above 45% before PLTL was implemented to below 20%. The withdrawal rate also decreased from above 25% to less than 10% during that same period.

The campus's success with conventional PLTL prompted peer leaders and investigators to question what more could be accomplished with the model. Excited about the possibilities of online PLTL workshops, two undergraduate students, Kevin Mauser and John Sours (who served as peer leaders and student coordinators for the face-to-face workshop), conspired to lobby Dr. Varma-Nelson until she agreed to support the development and testing of a working model. From the beginning they were the driving force behind cPLTL's development. With web conferencing software and equipment, they made it possible to adapt PLTL to a synchronous virtual environment.

Because of its rich culture of IT innovation and e-learning, IUPUI was well suited for the development of cPLTL. It had a well-established PLTL program in first-semester general chemistry. It is an urban research university with a diverse commuter student population, which was ideal for testing the model. Of its 30,000 students, 15% are minorities, 36% are 25 years or older, and 68% work off campus (*37*).

The development was executed by an interdisciplinary team of investigators with expertise in chemistry; instructional technology; and higher education research, evaluation, and theory. The project (http://cpltl.iupui.edu) represents a partnership between the CTL, the Department of Chemistry and Chemical Biology, and the University College at Indiana University Purdue University Indianapolis (IUPUI).

In 2010, NSF provided funding to develop cPLTL and study the conditions and tools required for enhanced cyber-learning via PLTL model (*NSF-DUE-0941978*). NSF support was critical for the development and testing of the model. IUPUI developers have now designed, implemented and tested a robust online platform to deliver PLTL using real-time chat, video conferencing, document sharing, and desktop sharing capabilities as shown in Figure 1.

Anatomy of cPLTL Workshops

Typically, six to eight students and one trained peer leader login to a web-conferencing meeting room. Once in the environment, each person shares his or her webcam, microphone and USB document camera. With prompts and guidance from the peer leader, the students work through problem sets, case studies, or other course-related content. The document camera share window allows students to observe each other's work, comment, and guide their peers. Workshop participants can also pair-off and enter virtual rooms to work in smaller groups, all while being observed and guided by the peer leader. Figure 1 depicts the basic structure of a cPLTL workshop.

The cPLTL virtual workshop takes advantage of common web conferencing service user interface components such as the following:

- *Participant's list* - shows the names of all people in the room at any time. This list allows the peer leader to see who enters or leaves the room during the session.

- *Audio/Video sharing window* - allows the peer leader and students to see and hear one another during the workshop session.
- *Chat window* - allows peer leaders to share links or instructions for activities. It may also be used as an alternate mode of communication in the event of a technical glitch (with headsets, microphones, or web-cameras).
- *Presentation window* - allows each participant to share his or her own work via their *document camera* while simultaneously viewing the work of every other student in the workshop. This format provides an environment in which students can engage in course material and complete the work collaboratively.
- *Two Cameras* - the key technology ingredient of cPLTL is the capability of using two cameras simultaneously. The document camera captures each student's work while their web-camera displays the real-time image of the student.

Figure 1. Basic structure of a cPLTL workshop. (see color insert)

Examining cPLTL's Effectiveness

It was well documented that PLTL workshops had significant positive impact on student achievement, attitudes, and persistence (*19*). However, moving the workshops to an online platform carried a number of concerns. It could present several new challenges for the leaders who facilitate the groups and have negative impact on student achievement. When IUPUI investigators implemented the cPLTL program, their primary question was "*What impact does cPLTL have on student outcomes—achievement and course completion?*" Since IUPUI's cPLTL pilot in 2009, the project's evaluation team has examined the educational outcomes of the online (cPLTL) and face-to-face (PLTL) workshops and compared participants' nationally normed *American Chemical Society's (ACS) First-Semester General Chemistry exam (2005)* scores; mean and end-of-course grades; interview, survey, observation, and discourse data to document the impact on student outcomes.

At IUPUI, PLTL workshops are mandatory for students who enroll in the general chemistry course; however, students self-select enrollment in the cyber and face-to-face workshops. From Fall 2009 thru Spring 2011, 181 first-semester general chemistry students (78 cPLTL and 103 PLTL students) participated in the IUPUI study. An analysis of their course grades and ACS exam scores showed those in cPLTL workshops performing at the same level as their peers in face-to-face workshops.

There were no statistically significant differences in the mean final grades earned in the general chemistry course. On average, students in the sample earned approximately a C+ grade in the course, with cPLTL students earning a mean grade of 2.37, and PLTL students earning a 2.30. Eighty-two percent of participants in cPLTL workshops received a final grade of C or better, compared to 78% of participants in PLTL workshops. There were also no statistically significant differences between the mean percentage score (m = 59) earned by cPLTL students and the score (m = 57) earned by PLTL students on the ACS exam, suggesting cPLTL was at least as effective as the traditional PLTL course.

After successfully developing and implementing the program at IUPUI, the team received additional support to expand the use of cPLTL at other institutions. The Next Generation Learning Challenge (NGLC) Wave I initiative (http://www.nextgenlearning.org/college-completion), coordinated by Educause (on behalf of the Bill and Melinda Gates and William and Flora Hewlett foundations), facilitated the scaling and dissemination of the project's research products by engaging the project team in conference, publication, and collaborative activities with institutions (supported by Wave I funding) focused on similar initiatives in improving STEM education. NGLC has also provided funding to scale up and study the transportability of cPLTL at other institutions and in other disciplines.

NGLC support has allowed Florida International University (FIU) and Purdue University (PU) to participate as cPLTL consortium members and implement cPLTL on their campuses. The model was implemented in introductory biology courses in Fall 2011 at FIU and in Spring 2012 at PU. The additional support allowed the cPLTL program to expand at the IUPUI campus as well.

When cPLTL was implemented (or expanded) at these institutions, FIU already had a well-established PLTL program. Purdue had no PLTL program, and therefore, no control group. Neither PLTL nor cPLTL was mandatory at FIU: Students self-selected to participate in the program(s). Purdue, however, did make cPLTL mandatory for a particular section during their implementation period. For each consortium institution, investigators analyzed *end-of-course* mean grades and the percentage of students earning A, B, or C grades and those earning D, F, or W grades. At each campus with a study and control group, there was no statistically significant difference in academic performance between the groups (*38*), indicating that cyber-students perform as well as students in the widely successful face-to-face workshop. Although there was no PLTL control group for PU, the *D, F, W* rate (15%) for their cPLTL students was slightly lower than the baseline *D, F, W* rate (18%) for the course.

Student Experience Surveys

IUPUI's student experience surveys indicated that both groups (cPLTL and PLTL students) were generally satisfied with the support they received from peer facilitators. Just over 80% of students in cPLTL and PLTL workshops agreed or strongly agreed that interactions with their peer leader helped to increase their understanding of the material. Similarly, students were at ease in their interactions with their peers in cyber and face-to-face workshop environments, with 83.3% of cPLTL participants and 81.5% of PLTL participants agreeing or strongly agreeing they were comfortable offering assistance to their classmates.

However, two survey items did show statistically significant differences between the perceptions of cPLTL and PLTL participants. While more than half of cPLTL participants agreed or strongly agreed that they enjoyed participating in the workshops (63.9%) and that their knowledge and understanding of the course material was a result of their participation in cPLTL workshops (65.7%), a larger portion of PLTL students enjoyed participation in the face-to-face workshops (73.6%) and felt their knowledge and understanding of the course material was a result of participating in PLTL workshops (83.2%). The evaluation team is now matching demographic data with participation, performance and survey data to further contextualize the findings and to determine the impact on low-income (Pell-eligible) student populations.

Focus Groups

In focus groups with peer leaders, investigators also found that the connections formed in cPLTL workshops may not translate into social relationships beyond the workshop as they do with PLTL students. For example, while students were friendly and comfortable in the online environment, some did not recognize members of their workshop group or engage with peer leaders when they encountered them in other settings on campus. This is a divergence from what has typically been found with participants in face-to-face PLTL workshops and raises questions about how the cyber-environment impacts the development of a "community" of learners. This finding requires further investigation.

Key Outcomes

The goals of the cPLTL project were to 1) test whether PLTL could be effectively used in an online environment, and 2) test whether it was transferrable to other institutions. Collectively, the observations of cPLTL workshops and analysis of course grades, standardized exam scores, and *D, F, W* rates have shown that cPLTL has the same impact as conventional PLTL. Since its implementation in 2010, cPLTL has been offered in 37 general chemistry workshop sections at IUPUI, 12 biology sections at Purdue University, and 16 biology sections at Florida International.

To date, all measures have shown cPLTL's positive impact on student learning in chemistry and biology. Its efficacy is further highlighted by its recent designation as a Sloan-C 2012 Effective Practice in Online and Blended Education award recipient (http://cpltl.iupui.edu/News). The results of the past year's multi-site evaluation have confirmed the cyber-workshops can be used effectively in STEM disciplines beyond chemistry and across a range of institutions.

This work demonstrates that cPLTL trains students in collaboration and leadership in an online environment, an important twenty first century skill. The work has expanded the cPLTL program and increased our knowledge of what works, what is necessary for a successful cPLTL program, as well as, what still requires tweaking.

Lessons Learned

There were a number of lessons learned over the course of the project. The team has since developed guidelines and best practices for cPLTL implementation (http://cpltl.iupui.edu/News). Included among these are recommendations for materials development, workshop training, technology use, and partnerships with students. A sampling of these issues are described below.

Materials Development

Immediate web access held both advantages and disadvantages. For a more balanced comparison, the materials used in cPLTL were identical to those used in PLTL. However, direct access to the wealth of information available on the web may, to some extent, diminish student motivation to solve complex problems through their own cognitive efforts. The workshop materials need to be challenging in a way that forces students to grapple with and weigh the feasibility of alternative solutions. New materials, better suited for the cyber-learning environment, need to be developed to challenge and motivate student learning.

Workshop Training

Training leaders for workshop participation has always been a hallmark of the PLTL model (*21*, *39*). It has become clear that students should also understand how their role in this environment will be different (e.g. requires active participation in discussion, collaborative problem-solving, and explaining concept, etc.) and be prepared to participate as a part of a community of learners. In addition, all cPLTL participants need to be trained with the technology, and on workshop expectations, and learning within a social context. With that goal, the development team implemented a new feature, *Workshop Zero*, in which students participate in a simulated workshop before their official workshops begin.

Technology Use

Similar to the dissemination of PLTL, developers articulated the workshop critical components to adopters, but were less prescriptive about how they should be interpreted and modified within the context of institutional culture, type, and students demographics. This resulted in unique cPLTL implementations at different institutions. Investigators understood that adaptation varies from location to location and is based on what is needed within the adopter's local context, but they did not clearly distinguish program aspects that were essential for success in the "online" learning environment. For instance, some disciplines require more discussion than display of visual aids and models. This prompted users to forego use of the document camera. Limited bandwidth also dissuaded adopters from using real-time participant images during workshops. However, the real-time images of participants provide important information to leaders and other group members. For example, body language and facial expressions help participants and leaders identify when their peers are struggling with the material and need additional support. It also indicates to leaders when students are distracted, have technical difficulties, or are off-task.

Student Enrollment Decisions

At IUPUI, we observed lower than expected enrollment in the cPLTL workshop. To better understand why, we administered an adhoc survey to IUPUI cyber and face-to-face students regarding their workshop preferences. In Fall 2011 through Spring 2012, we asked students what was most important in choosing a cyber or face-to-face workshop and whether they were aware of the online option.

The data showed that it was important for both groups to have a workshop that fit their schedule, according to 93.6% of cPLTL students and 72.7% of PLTL students. The most significant differences between the two groups were that cPLTL students (58.7%) wanted to avoid the commute, and only 14.7% of them preferred face-to-face learning; while students (85.9%) in the face-to-face workshops preferred face-to-face learning and 76.4% of them favored taking courses on-campus. Among the face-to-face PLTL survey respondents, more than half (53.3%) were not aware of the cyber version of the workshop.

IUPUI students typically learned about the PLTL workshop requirement after they enrolled in the general chemistry course. The same was true for FIU. The data indicated that more needed to be done to publicize and improve awareness of the cPLTL program among entering students before they enrolled in courses. The recruitment arm of the development team has increased outreach efforts to inform entering students about the program before the start of the semester. The training and orientation teams have also made changes to the cPLTL orientation schedule. Previously, cPLTL orientation was required for participation in the cPLTL workshop program, but offered little flexibility in scheduling options for prospective participants.

Partnerships with Peer Leaders

Having strong relationships with students (especially peer leaders) enhance the likelihood of success. Peer leaders are trained to lead. They give feedback to faculty on how the workshops are going, how the materials are working, and what students understood or misunderstood in the lecture. In this role, they become partners with faculty in the cPLTL model. In each implementation outlined above, partnerships with students were crucial. Peer leaders support the sustainability of the program and are very effective in convincing other faculty to adopt cPLTL in their courses, training and recruiting new leaders, creating supportive environments for students, as well as, proposing and developing innovations with PLTL. It is not surprising then that the development of cPLTL was catalyzed and spearheaded by two undergraduate students Kevin Mauser (now a biomedical engineer) and John Sours (currently a medical student) who had served as peer leaders in the PLTL program already flourishing when Varma-Nelson arrived at IUPUI. Their contributions to the development of cPLTL were so significant that they are the first two authors of our first paper (*40*).

Sustainability

After only one year, there is evidence of cPLTL's sustainability at other institutions. Typically, the primary cost associated with sustaining a PLTL program is compensating peer leaders for facilitating workshops. However, as with PLTL, institutions have developed ways of addressing this issue so that it has no negative impact on cPLTL's sustainability. For example, Purdue undergraduate peer leaders now receive course credit for their work, and this leadership development course is now part of the biology curriculum for science majors. Their students who participated as paid peer leaders during the NGLC grant period are now "enthusiastically volunteering to return as peer leaders," eliminating the need for departmental funds for leader compensation.

Similarly, Purdue graduate students (who piloted cPLTL in a graduate course) are volunteering to participate as cPLTL leaders in the graduate level course because it enhances their professional development and teaching portfolio. The cPLTL workshops further help the department meet requirements for the *Undergraduate Outcomes-Based Curriculum and Administration & Oversight Structure*, as the PLTL component helps students demonstrate critical thinking, ethical reasoning, leadership and teamwork, quantitative reasoning, integrative knowledge, written communication, information literacy, and oral communication. As a result, both Purdue courses are currently being supported by the biology department.

FIU has an extensive PLTL program and traditionally has not paid peer leaders for workshop facilitation. Instead, FIU uses other types of incentives to compensate peer leaders. For example, their peer leaders may receive credit for service learning or leadership or may receive letters of recommendation for graduate programs or medical school applications. FIU cPLTL coordinators expect the option to facilitate the workshops remotely will be a welcomed incentive for their peer leaders.

Because IUPUI has a PLTL program that is supported by the institution, peer leaders simply choose to facilitate online or in face-to-face workshops. Essentially, there is no added cost to compensate peer leaders. Similarly, because IUPUI has contracted with Adobe to offer the Connect service, no additional cost is incurred for the cPLTL implementation, thus promoting cPLTL's sustainability. However, other options are being explored, specifically through IUPUI's piloting of the learning management system (LMS) tool, Canvas, which contains a built-in web-conferencing component. cPLTL investigators will test Canvas' viability for supporting an implementation of cPLTL.

Broader Impacts

The project's products (database of recorded workshops) have provided a new way to gauge student progress in understanding course content, assess peer leader development needs, devise and enhance leader training programs and initiatives, and engage faculty in reflective practice. Previously, obtaining PLTL data on student interactions in the workshops was tedious work which required invasive methods (i.e., manual audio video recording). With cPLTL it is now possible to automatically capture all chat sessions, written collaboration, voice recordings, and video in a non-intrusive manner. This allows PLTL data analysis at a level of detail that has not been possible in the past. And, as a result of the recorded workshop sessions, there is now an extensive cPLTL database which has the potential to keep education researchers occupied for several years. The recordings also allow faculty to:

- monitor, in a non-intrusive way, how effectively peer leaders facilitate group discussions and how well they convey accurate information during those interactions;
- gauge the efficacy of their leader training initiatives and to better align pedagogical and content training with peer leader development needs;
- inform their practice as they provided unfiltered insight on students' content knowledge (what students are or are not getting from lectures); and
- gauge whether they have achieved goals for student learning, refine how they convey information in lectures, and restructure lectures and develop course materials to better suit student learning needs.

The current findings expand the knowledge base on best practices in STEM education and aid STEM faculty and administrators in understanding how to better support students' online learning needs, develop suitable instructional methods, and use technologies that are more effective in helping students engage in, understand, and apply course (chemistry/biology) material. More specifically it informs educators about:

- How students interact with their peers, course materials, educational resources, and the technology

- What types of educational resources students use/access to understand content from lectures and workshops
- How the technology facilitates or impedes students' understanding of content
- Student support needs and provisions over the course of the workshop

Future Directions

IUPUI has now offered cPLTL for seven semesters in general chemistry, and has facilitated cPLTL offerings in biology at Purdue University and Florida International University. As the team moves forward, it is exploring options and developing new course materials for cPLTL workshops to enhance engagement and further challenge students who have access to new resources in problem-solving activities. The team is expanding cPLTL adoption to disciplines beyond chemistry and biology (at IUPUI) and to broader student populations. To this end, Sinclair Community College of Dayton, Ohio is now planning cPLTL implementation. The team is also further testing other hardware and web conferencing tools (e.g., Google Hangouts, BigBlueButton, and OpenMeeting) at other institutions.

Currently, institutions that already have access to web conferencing software can readily adopt cPLTL. Those who do not may choose from a host of other web conferencing resources. These institutions may experiment with the cPLTL model and investigate free and/or open-source web conferencing tools.

With the exception of a USB document camera ($60) and microphone/ headset ($10), it is likely that most learners enrolled in an online course will have a webcam ($15) and computer (see Figure 2.).

Figure 2. cPLTL technology set. (see color insert)

Document cameras are important for disciplines such as chemistry, mathematics and physics where problem solving and sharing in the process with others is important, but may not be as important in fields like biology or psychology. Participants can use a webcam or a tablet input device; and for some instructional content, document cameras may not be necessary. Purchased together, the three items total $85, which is less than many required textbooks.

Potential Role of cPLTL in MOOCs

As interest and investment in Massive Open Online Courses (MOOCs) increases, so too does the need to improve the quality and structure of online learning. The promise of MOOCs lies in their ability to extend access to course content to broader populations. However, despite their potential, MOOCs are closely scrutinized in regards to their:

- withdrawal rates--as only 5-14% of MOOC participants complete or enroll with the intent to complete courses (*41*);
- level of active engagement and deep learning;
- level of feedback from more informed mentors;
- level of attention to the needs of novice and individual learners; and
- quality and level of personal interaction and sense of community.

Existing challenges in these areas limit the scope of student learning and development that virtual environments provide. However, cPLTL could be a solution to several of the issues.

The high attrition rate in MOOCs and online courses in general suggests more needs to be done to improve the sense of community and *mattering* among students (*33*, *42*), as the strength of their group/social cohesion greatly impacts their institutional commitment and their intentions to persist (*42*).

Creating group/social cohesion may not be so improbable for MOOCs. Some students enrolled in MOOC's have begun to take part in MOOC meet-ups, informal in-person course discussion groups. These groups take form based on the geographical location of the participants. While they may, to some degree, offer more meaningful interaction and engagement, they may be missing a key element, an informed mentor. However, MOOC providers may be able to capitalize on the construct of MOOC meet-ups by offering them online, making them more structured (with appropriate materials), and by identifying local mentors (within similar time zones) interested in facilitating the groups. These mentors could be screened, trained, and certified by virtual means (*e.g.*, electronic badges). With such a system, the virtual and synchronous nature of cPLTL (with peer interaction, engagement, mentors, and the supportive learning community embedded) makes the workshops an attractive option for traditional online courses and MOOCs.

Acknowledgments

Finally, we gratefully acknowledge the support of the project's sponsors: IUPUI, NSF (*DUE-0941978*), and the NGLC Wave I and Follow-On Initiatives (EDUCAUSE on behalf of the Bill and Melinda Gates and William and Flora Hewlett foundations). We also acknowledge the contributions of John Sours, Kevin Mauser, and the many other peer leaders; lecturers, Lin Zhu, Hongqiu Zhao, and Gina Londino; instructional technologists, Thomas Janke, Randy Newbrough, and Lorie Shuck; the external evaluator, Donald Wink; CUME research associates; university partners, Nancy Pelaez (PU) and Thomas Pitzer

(FIU); and the advisory board, Chris Bauer, Thomas Higgins, Jeffery Moore, and David Yaron.

References

1. Vygotsky, L. The Collected Works of L. S. Vygotsky. Vol. 2: The Fundamentals of Defectology. In *Abnormal Psychology and Learning Disabilities*; Rieber, R., Carton, A., Eds.; Plenum Press: New York, 1993.
2. Johnson, D. W; Johnson, R. T. *Educ. Res.* **2009**, *38*, 365–379.
3. Goldstein, L. *Am. Educ. Res. J.* **1999**, *36*, 647–673.
4. Carnevale, A. P.; Smith, N.; Melton, M. *STEM: Science Technology Engineering Mathematics*; Georgetown University Center on Education and the Workforce: Washington, DC, 2011.
5. Gosser, D. *J. Peer-Led Team Learn.* **2011**, *14*.
6. Gosser, D. K.; Roth, V. *J. Chem. Educ.* **1998**, *75*, 185–188.
7. Gosser, D.; Roth, V.; Gafney, L.; Kampmeier, J. A.; Strozak, V.; Varma-Nelson, P.; Radel, S.; Weiner, M. *Chem. Educ.* **1996**, *1*, 576.
8. Varma-Nelson, P.; Coppola, B. Team Learning. In *The Chemists' Guide to Effective Teaching*; Pienta, N., Cooper, M., Greenbowe, T., Eds.; Prentice Hall: Upper Saddle River, NJ, 2004; pp 155−169.
9. Historically Black Colleges & Universities – Undergraduate Program (HBCU-UP). Morehouse College Receives $2.4 Million for HBCU-UP. *HBCU Undergraduate Prog. Newsletter*, April 2007, 1.
10. DiCecco, J.; Silva, S.; Coughlin, J.; Swaby, T., Wu, J.; Sun, Y.; et al. *Experiential Learning in Neurophysiology for Undergraduate Biomedical Engineering*; Department of Biological Sciences, University of Rhode: Kingston, RI, 2004.
11. Preszler, R. *CBE-Life Sci. Educ.* **2009**, 182–192.
12. Flores, B.; Becvar, J.; Darnell, A.; Knaust, H.; Lopez, J.; Tinajero, J. *Am. Soc. Eng. Educ.* **2010**.
13. Hooker, D. *Multicultural Perspect.* **2011**, *13*, 220–226.
14. Murray, J. D. Peer Learning and Its Applications to Undergraduate Psychology Instruction. In *Promoting Student Engagement*; Miller, R. L., Amsel, E., Kowalewski, B. M., Beins, B. C., Keith, K. D., Peden B. F., Eds.; 2011; pp 166−169.
15. Horwitz, S.; Rodger, S. *Using Peer-Led Team Learning To Increase Participation and Success of Under-Represented Groups in Introductory Computer Science*; SIGCSE: Chattanooga, TN, 2009.
16. Powell, R. M.; Murphy, C.; Cannon, A.; Gordon J.; Ramachandran, A. *Emerging Scholars Program: A PLTL-CS Program That Increases Recruitment and Retention of Women in the Major*; Department of Computer & Information Science (CIS), University of Pennsylvania: Philadelphia, PA, 2012.
17. White, P.; Rowland, A.; Pesis-Katz, I. *J. Nurs. Educ.* **2012**, *51*, 471–475.
18. Dobson, G.; Frye, R.; Mantena, R. *Am. J. Bus. Educ.* **2013**, *6*.

19. Gosser, D. K.; Kampmeier, J. A.; Varma-Nelson, P. *J. Chem. Educ.* **2010**, *87*, 374–380.
20. Gosser, D.; Cracolice, M.; Kampmeier, J.; Roth, V.; Strozak, V.; Varma-Nelson, P. *Peer-Led Team Learning: A Guidebook*; Prentice Hall: Upper Saddle River, NJ, 2001.
21. Gafney, L.; Varma-Nelson, P. *Innovations in Science Education and Technology: Peer-Led Team Learning: Evaluation, Dissemination, and Institutionalization of a College Level Initiative*; Springer: Weston, MA, 2008; Vol. 16.
22. Akinyele, A. *Chem. Educ.* **2010**, *15*, 353–360.
23. Báez-Galib, R.; Colón-Cruz, H.; Resto, W.; Rubin, M. *J. Chem. Educ.* **2005**, *82*, 1859–1863.
24. Hockings, S. C.; DeAngelis, K. A.; Frey, R. F. *J. Chem. Educ.* **2008**, *85*, 990–996.
25. Lewis, S. E.; Lewis, J. E. *J. Chem. Educ.* **2005**, *82*, 135.
26. Lyon, D. C.; Lagowski, J. J. *J. Chem. Educ.* **2008**, *85* (11), 1571–1576.
27. Tien, L.; Roth, V.; Kampmeier, J. *J. Res. Sci. Teach.* **2002**, *39*, 606–632.
28. Wamser, C. *J. Chem. Educ.* **2006**, *83*, 1562–1566.
29. Rankin, S.; Reason, R. *J. Coll. Stud. Dev.* **2005**, *46*, 43–61.
30. Allen, I. E.; Seaman, J. *Staying the Course: Online Education in the United States*; The Sloan Consortium: Needham, MA, 2008.
31. State Higher Education Executive Officers (SHEEO). *State Higher Education Finance FY 2012*; SHEEO: Boulder, CO, 2013.
32. Allen, B. *Community Services Catalyst*, 1993, *23*, No. 3. http://scholar.lib.vt.edu/ejournals/CATALYST/V23N3/allen.html (accessed July 3, 2013).
33. Jacoby, B.; Garland, J. *Coll. Stud. Retention* **2004**, *6*.
34. Pascarella, E. T.; Terenzini, P. T. *How College Affects Students*; Jossey-Bass: San Francisco, 1991.
35. Astin, A. W. *What Matters in College: Four Critical Years Revisited*; Jossey-Bass: San Francisco, 1993.
36. National Survey of Student Engagement. *Experiences That Matter: Enhancing Student Learning and Success*; Indiana University Center for Postsecondary Research: Bloomington, IN, 2007.
37. IUPUI Office of Communications and Marketing. *Indiana University-Pudue University, Indianapolis, Quick Facts 2009*. http://www.iupui.edu/about/facts.html (accessed April 7, 2013).
38. Smith, J.; Banks, J.; Zhu, L.; Wilson, S.;Varma-Nelson, P. *Evaluating Cyber Peer-Led Team Learning (cPLTL)*; 2013, manuscript in preparation.
39. Roth, V.; Goldstein, E.; Marcus, G. *Peer-Led Team Learning: A Handbook for Team Leaders*; Prentice Hall: Upper Saddle River, NJ, 2001.
40. Mauser, K.; Sours, J.; Banks, J. V.; Newbrough, R.; Janke, T.; Shuck, L.; Zhu, L.; Ammerman, G.; Varma-Nelson, P. *Educause Rev. Online* **2011**, *34*.
41. Nicoara, E. S. *The Impact of Massive Online Open Courses in Academic Environments*; Conference Proceedings of e-Learning and Software for Education, Bucharest, April 2013; pp 644–649.
42. Hausmann, L. R.; Schoefield, J. W.; Woods, R. L. *Res. Higher Educ.* **2007**, 48.

Chapter 8

Working To Build a Chemical Education Practice

Donald J. Wink,[*,1] Sharon Fetzer Gislason,[1] and Julie Ellefson[2]

[1]Department of Chemistry, Learning Sciences Research Institute, University of Illinois at Chicago, 845 W. Taylor Street, Chicago, Illinois 60607
[2]Math and Science Division, Harper College, 1200 W. Algonquin Road, Palatine, Illinois 60067
[*]E-mail: dwink@uic.edu

Three chemical educators, working with others in a large metropolitan area, have followed a trajectory whereby they built a multi-faceted program of chemical education practice. This followed a trajectory of development that was done in part through three particular projects, one focused on course development, one on lab materials development, and a third on curriculum development. The three projects are described according to common themes that played out in very different ways: collaboration, interdisciplinarity, student-focus, and dissemination. But attending to these themes, long term development of sustainable programs occurred, with impacts at the individual, institutional, regional, and national levels.

This paper is written by three chemical education faculty who have worked together and with others, especially in the Chicago area, to forge an interdisciplinary community in support of learning in college chemistry. During this time, they followed a trajectory that took these individuals, with no formal chemical education training, to a point where they, their coworkers, and their students benefit from having a substantial, and enduring, chemical education community in their area. At the same time, the products of their work—textbooks, curricula, and research papers—also affect chemical education more broadly.

© 2013 American Chemical Society

In no small part, their work has been done with the support of the National Science Foundation, especially the Division of Undergraduate Education. And they have benefited frequently by reflecting on the themes suggested by DUE and DUE-funded projects.

This trajectory occurred at multiple levels and according to four themes—collaboration, interdisciplinarity, focus on students, and a goal of dissemination—that provided the framework for their work. After reviewing these themes, they then briefly describe how their own personal trajectories brought them to the point in 1992-3 when they first started to work together. The bulk of the chapter then discusses how their work was done in the context of three very different projects. They cap this by reviewing where things stand today, at this point in the trajectory of themselves, their projects, their institutions, and their regional community. Finally the authors synthesize their findings into a set of conclusions for developing a collaborative practice of chemical education that can be applied in many other potential projects.

Four Themes as a Framework

The work described in this chapter involves three very different kinds of intervention: an entire course design, a set of new laboratory experiments, and the chemistry part of a multi-course program. But all three shared much more than their mutual work. These projects all had four essential themes: interdisciplinarity, collaboration, a student-focus, and readily disseminated products. Interestingly, all four themes are also mentioned in Program Solicitations from the Division of Undergraduate Education of the National Science Foundation, including the 1991solicitation for the *Undergraduate Course and Curriculum Development Program* that was the basis of their earliest program designs (*1*).

Interdisciplinarity, in this work, meant seeking knowledge as fundamentally connected. This means, for example, that it is vital to understand not simply math and chemistry but also how math is used in chemistry, and how chemistry can serve students in understanding math. It is also crucial that these connections be made explicit. The interdisciplinary character of the work does not just come from the authors. Here, the *kind* of interdisciplinary work they engaged in was, in part, defined by the diverse disciplinary goals the students had for *their* learning. This led them to address the question "What good is chemistry?" for professions and disciplines as varied as engineering, nursing, and elementary school teaching.

Collaboration took a variety of forms in their work as they strove for particular elements that made the work that of peers, not of a hierarchy. This began, quite simply, with an effort to find coworkers in similar settings who would be able to broaden the perspectives used for these projects. Since the authors saw peer relationships as vital, it also meant that they tried to share responsibility and benefits equally, whether in co-authorship or in teaching the same material in different settings. Active feedback was central throughout their work, including listening carefully to what they could learn from each other.

These projects were informed by a **student-focus** at many levels. They were aiming to generate materials and curricula that would support students in their actual learning trajectories as noted earlier. This required them to inquire about the type of conceptual foundations students would need, about the sort of futures they had in store for them, and finally about the kind of interests they had. In practice this also meant they had to do a lot of listening: to pay attention to what was being said by others about what students should and could learn in anticipating their future professional lives. They also needed to hear what was being said by the students themselves as they described learning. Throughout, the authors also realized that they were receiving as much from the students as they were giving to them.

Finally, all of these projects had a plan for **dissemination** as part of their initial design. At first this was done in a formative way by sharing materials and ideas broadly so as to gain additional insight. The final goal was verbal and written publication of their results for wider inspection and application in settings where they were not active participants.

Starting a Trajectory

In a book about trajectories, it is important to note starting and end points, since they bookend the paths the three authors have taken. Here, a description of the authors 'state' in 1992 is discussed; later, briefer endpoints—where they are now—are presented.

There are many odd things about the path Wink followed to becoming a chemical education faculty member. This begins with the fact that he had no formal training in education and, prior to 1992, no informal training either. Rather, his background was entirely appropriate to that of a conventional synthetic inorganic chemist, with publications from college and graduate school focused on making highly air-sensitive compounds that were interesting in and of themselves—generally not for any "purpose" (*2*, *3*). This was enough to get him a position as a junior faculty member, where he continued to work in inorganic chemistry. During that time, he also worked on two different education-related projects: a change in a physical chemistry lab course using some of the very first funds available in NSF's *Instrumentation and Laboratory Improvement* program and in general chemistry. The lab course contained elements of "project based learning." But the general chemistry work was entirely from a perspective of traditional lecture. Both projects resulted in publications in the *Journal of Chemical Education* but, when denied tenure on the basis of his 'conventional' chemistry work, he was able to grab the brass ring of a position (remarkably, with tenure) at UIC in 1992.

His background as a conventional chemist gave Wink three things that were critical to his approach in chemical education. First, he had to be innovative on the basis of what had been done before, that is, to root his work in the literature of the field. Second, he had to ensure that the work could be recognized by extramural funding and publications, the *lingua franca* of demonstrating excellence in university work. And, third, he knew he needed a 'team.' But, since his new department did not have a functioning area of work in chemical education, he recognized that this team had to be developed with peers, not graduate students.

The third component of his approach was one that brought him together with Gislason (then Fetzer) and Ellefson. Fetzer Gislason and he met in his first weeks at UIC—she had just finished her PhD and had begun an appointment as a lecturer. With teaching as the focus of her work and her background, she was a natural collaborator. Together with faculty from math they generated the "MATCH" proposal (discussed later). But even then they knew to go beyond the boundaries of UIC, so in May, 1993 Wink picked up the phone and called Ellefson at Harper College to find a site for dissemination.

Fetzer Gislason began her teaching career at UIC following a mid-life change of direction in 1986. Before that time she was certified in secondary education, taught high school chemistry for seven years, married, and started a family. She planned to return to teaching when the youngest child started school. However, it turned out that she was a stay-at-home mom for almost 15 years, including 2 years spent at a local college to see whether or not she could survive going to graduate school while her four children were still in grammar school. Unfortunately, most of the universities she considered at that time catered to the working crowd, meeting on nights and weekends. Fetzer Gislason was seeking a program that met during regular school hours so she would not lose unnecessary time with her children. UIC, she discovered, offered graduate courses during the week so she took a deep breath, enrolled in their chemistry graduate program in 1986 and was on her way.

In addition to required classes, Fetzer Gislason worked as a teaching assistant for freshman-level chemistry courses for the first two years of graduate school. Later she was appointed the "head" TA whose chief responsibility was to supervise TAs in the general chemistry lab sections. One of her duties was to meet weekly with the director of undergraduate education, Wade Freeman, to review and edit pre-existing TA lab notes which she then retyped, copied, and distributed to the general chemistry TAs. On a daily basis she visited the chemistry labs and discussion sections led by these TAs, found replacements for those who could not make their sections, and generally gave assistance as needed.

After receiving her Ph.D. in January 1991, Fetzer Gislason began her teaching career at UIC. She was assigned to teach analytical chemistry during the 1991 summer session and then she continued on to teach preparatory chemistry (prep-chem) and general chemistry I during both semesters of the 1991-92 school year. These two courses are intimately related for incoming students with weak backgrounds in math and little or no chemistry knowledge. Achieving a grade of C or better in prep-chem allows them to enroll in general chemistry, a course that is required by 25 majors at UIC. These majors include but are not limited to engineering, pre-med, chemistry, biology, nursing, pharmacy and other health-related fields. Because most students directly follow the sequence from prep-chem to general chemistry and because Fetzer Gislason was getting many of the same students for both semesters, she was able to assess how well the preparatory class was helping her students to succeed in general chemistry.

She was dismayed to find that much of the "wonderful" teaching she thought she had done for the prep-chem students was not retained by almost 30% of those who enrolled in general chemistry the following semester. Students who had to repeat either prep-chem or general chemistry were effectually stopped at this point in their education from further progression toward their educational goals. This concerned Fetzer Gislason greatly. Thus when a newly hired chemical educator, Wink, joined the UIC faculty in 1992, Fetzer Gislason met him with the plea, "Our prep-chem course is *not* preparing students for general chemistry. Help!"

Ellefson gained her first full-time teaching experience during the final semester of her Master's degree program in chemistry. One of the chemistry teachers at a girls' Catholic high school had to leave mid-year. While she had no formal educational training, she had served as a teaching assistant and knew she wanted to pursue a career in education so Ellefson applied and was accepted for the position. Despite a grueling schedule teaching full-time and completing her research and writing her thesis, Ellefson realized that her decision to teach was the right one as her interactions with her students motivated her tremendously that semester.

The following year Ellefson was hired to teach high school chemistry, physics, and physical science as half of the science department at a small parochial high school. Her students had very different backgrounds and interests and ranged in age from freshmen to seniors. Some days Ellefson wished she could use a strategy the biology teacher employed to help her students stay focused – hold and pet one of the guinea pigs. Since that was not an option, Ellefson incorporated demonstrations and group activities into her classes as she began shaping her student-focused teaching philosophy. Because she was not yet certified to teach, she began her formal educational training by enrolling in education courses.

After three years at the high school, Ellefson was hired as a chemistry instructor at Harper College in August 1988, where she initially taught general and preparatory chemistry. Although general chemistry has both a math and chemistry prerequisite, Ellefson found these classes populated with students with quite diverse backgrounds and varied levels of preparation, who, when asked almost always said they were taking the class because it was required for their intended major. On her second peer evaluation, her colleagues noted she had "a slight reputation with students of being rather demanding of students." She knew she wanted to maintain high standards, but that she also wanted to provide the support her students needed to succeed. Additionally, she wanted to help her students not only to learn to appreciate chemistry but to understand its relevance to their future careers and their lives.

Ellefson continued her formal educational training by earning a Master's in Education in 1993; her projects for the courses primarily focused on general chemistry issues. Thus she was involved in small scale general chemistry reform efforts throughout her early years at Harper including alternative methods of assessment, inquiry-based labs and activities, cooperative group work, and projects dealing with "real world" applications of chemical principles. These efforts were informed, in part, by the 1989 National Science Foundation Disciplinary Workshop on Undergraduate Education's call for the implementation of broad curricular reform in chemical education and the American Chemical Society Division of Chemical Education's Task Force formed in response to address the reform of general chemistry. However, it was not until 1993, when Donald Wink called and invited the Chemistry Department at Harper to participate in the MATCH program, that Ellefson was presented with the opportunity and developed the confidence to try to make a difference in both her own classroom and more broadly in the chemical education community.

The MATCH Program: A Combined Math and Chemistry Curriculum

When Wink and Fetzer Gislason began working on a revised version of prep-chem in 1992 they realized that a simple revision of existing materials would not accomplish their goal. At about this time, Wink attended a meeting for a wide-ranging *Alliance for Minority Participation* (AMP) program at NSF and met John Baldwin of UIC's Department of Mathematics, Statistics, and Computer Science. Baldwin put forth the idea that UIC's earliest developmental / remedial courses should support the student-learning of math. To his way of thinking, math informs chemistry and deep learning of chemistry relies on a fundamental understanding of mathematical principles.

Wink immediately saw the relevance of this idea to UIC's prep-chem class revision. Students who place into a UIC prep-chem class typically also place into a corresponding intermediate algebra class, i.e., many students began their college career with a double deficiency in math and in chemistry. In recognition of this, most prep-chem books seek to remedy the problem by placing a general math review in the first chapter. About a third of the students, unfortunately, do not possess the skills necessary to take advantage of such an early math review as these skills are taught much later in their math course. In addition, the math is rarely connected to how the students learned math in the first place—not that Wink or Fetzer Gislason knew much of what was being done in detail in math courses. So, they decided to confer with a math instructor currently teaching intermediate algebra to better understand how and where the two courses overlapped. Enter Sheila McNicholas. At the time, McNicholas was a lecturer in UIC's Department of Math Statistics and Computer Science where she taught introductory algebra and calculus.

McNicholas quickly informed them that they taught the math component completely wrong. The silence that followed was deadening until Wink asked her, "How so?" Wink and Fetzer Gislason were told that the "cutesy" manner that science teachers sometimes adopted in order to make math "easier" only made it more incomprehensible to students. Instead, McNicholas insisted, math is better taught using correct mathematical terms and expressions that the students can connect to previously learned mathematical concepts and skills.

After further discussion the three identified the problem as two fold: a lack of the problem solving skills for chemistry students and a lack of real applications for math students. There followed intense cross-questioning, defensive posturing, and multiple suggestions until someone said, "Hmmm. What if chemistry students received the relevant math instruction right when they needed it? What if math students got some real life applications from chemistry? What if we taught prep-chem and intermediate math as a single course?" The direction now seemed clear: McNicholas joined with Wink and Fetzer Gislason on a pathway to make this happen.

The framework of a possible interdisciplinary course slowly (and, for them, surprisingly) emerged from this collaboration. Instead of merely seeking insight from a math professional, their focus shifted toward the possibility of combining their skills, experience, and expertise in the production of a combined course. Together they applied to NSF for a planning and implementation grant. As this would be a math-chemistry project, it was aptly named MATCH, an acronym combining the names of the two disciplines involved. An exciting and daunting possibility lay before them as they proceeded into the planning phase.

The team began by compiling a list of chemistry topics that required some math understanding and then pairing as many topics as possible with the corresponding topic from intermediate algebra. Many, but not all, of the topics could be matched in this manner. Math and chemistry topics whose content did not overlap were taught independently of each other. The team decided that math would be introduced into prep-chem, as needed, while maintaining strict adherence to correct mathematical language and procedures. These math portions were authored by McNicholas. To get this to work, Wink and Fetzer Gislason had to delay the introduction of chemical principles that were dependent on math. Instead of beginning with a math review, they began their course with non-mathematical topics that would enable students to begin talking and thinking about chemistry. For example, during the first weeks of class students learned about molecular concepts, periodicity, linear equations, and the use of formulas in math and chemistry. Wink and Fetzer Gislason hoped to provide solid chemical concepts and skills through which students could learn some basic chemistry concepts and relationships. They wanted students to start learning the vocabulary and to be able to "speak" chemistry.

The MATCH grant was funded for a three year period beginning in June 1994 and the writing began in earnest that summer. Although the aim was to produce an integrated text, it was decided initially to write separate math and chemistry scripts so that each revised work could be checked for accuracy and for fulfillment of departmental requirements. The idea was to team-teach the course, each teacher stepping in to teach as needed. During this developmental phase both teachers were always present during class. As a benefit of the team-teaching approach, Fetzer Gislason was able to assess student difficulties and act as a mediator between struggling students and McNicholas. On the other hand, McNicholas, who remembered very little chemistry, could tell Fetzer Gislason when she had "assumed too much" and lost students at a crucial point in lecture because she had not made the intellectual connection for them.

Students received a syllabus that contained chemistry topics and pages matched with corresponding math topics and pages. The course included three main components: (1) Lectures four times a week for a total of 250 minutes; (2) Discussion sections twice a week for a total of 200 minutes; and (3) Laboratory sections once a week for 110 minutes. Although the lecture was mainly traditional, students were encouraged to work together to solve selected problems that tested their understanding and mastery of new material. Cooperative learning techniques were also used during their discussion sections when students gathered in small study groups to work through math/chemistry worksheets that used and extended information from lecture. With the addition of the lab experiments, the class consisted of 9 credit hours, 5 credit hours for math and 4 for chemistry. Students were kept within the same group and with the same TA for discussion and lab sessions. Peer mentors, supported through the AMP grant led by Chicago State University, were also used in the discussion sections. The result was close student-student bonding, student-faculty bonding, and more talking about chemistry and math among students. While all quizzes and exams integrated math-chemistry topics, grading was kept within each department for easier tracking. The full development of the "MATCH PROGRAM" was as follows:

- Summer 1994-a simple pairing of component course sections
- 1994-1995-writing a "lecture" textbook. During the first year of the grant, all student materials were photocopied and inserted into Class Binders which were provided. Later material was generated as needed, including lecture demonstrations that were revised into laboratory experiments.
- 1995-1996-writing of a separated chapter textbook and dissemination to Harper College for class testing.
- 1996-1997-writing a fully integrated Preparatory Chemistry and Intermediate Algebra text. This integrated version was disseminated to Chicago State for class testing and the program was institutionalized at UIC.

In addition to process of developing and teaching the materials, they also engaged in a process of studying the impact of the program in a more rigorous way. Working with Barbara Zusman in UIC's office of Data Resources, they developed surveys and student tracking methods that allowed for the program to be evaluated using quasi-experimental methods. The MATCH program, as noted, involved students who registered at UIC for two courses—special sections of intermediate algebra and prep chem. But not all students who could register for these special sections did so, creating a comparison group that were taking these courses in a traditional 'disconnected' manner. With Zusman's surveys tracking student backgrounds and attitudes, they confirmed that the groups were equivalent, permitting research data, and well-grounded claims of success, to be developed and published. Part of the personal trajectory for them was when they began presenting these research results at conferences and, eventually, in an article in *The Journal of Chemical Education*, assisted also by Rob Mebane, a faculty member who had a visiting appointment at UIC (*4*).

The dissemination of the project to Harper College was where Wink and Fetzer Gislason forged their connection to Ellefson. The dissemination was more than just a process of handing over materials. Rather, they all three saw it as a way for the MATCH program to gain the insight of a skilled chemical educator working in an environment that was very relevant to the developmental / preparatory goals of the MATCH program. It also allowed the authors to develop a sense of listening collaboratively across institutions, extending the process they had started with the math department and McNicholas.

From MATCH to *The Practice of Chemistry*

Cognizant of their long term goal of disseminating any materials they developed, Wink, McNicholas, and Fetzer Gislason signed an Agreement for Publication contract with W H. Freeman and Company for the complete revision of the MATCH program in March 1998. The new book titled *The Practice of Chemistry* (*POC*) was published in 2002 (*5*). *POC* followed the instructional strategy of MATCH; to introduce students to chemistry as it is communicated verbally, symbolically, and quantitatively while including a rigorous treatment of math in problem solving. The book was divided into three main parts: 1) Characterizing chemical substances and chemical reactions 2) Chemical quantities and 3) Chemical systems.

Although McNicholas alone wrote the math materials, Wink and Fetzer Gislason shared the writing of the chemistry portion. They began by each writing alternate chapters, then editing each others' chapter before sending it off to their editor at Freeman. The editing process was long. Each time the manuscript was sent back to Wink and Fetzer Gislason, the editing roles were reversed until both had written parts of every chapter and the manuscript had the same "tone" throughout. Although their editors strongly supported their ideas, they also wanted special features within each chapter that would unify it as a book. Thus the math sections were reformatted as "Making it work with mathematics" sections and inserted within the appropriate chapter. The practical chemistry connections described in MATCH were reformatted simply as section "Practicals."

For its time, POC contained some unique media features, electronic tools specifically created for, and closely integrated with, the text. The Web Tutor tool walked students through tough problems and provided them feedback. The Web Animator tool provided animated, three-dimensional molecules so students could visualize chemistry. The Web Practice tool allowed students to review chapter material using flashcards and online quizzing.

The word "practice" reflects the authors' ideas of what the *Practice of Chemistry* includes. An understanding of chemistry is required by many professionals whose daily work incorporates some chemistry. The word "practice" also indicates that study skills must be done repeatedly in order to be able to take on more difficult material in the future. Lastly 'practice" is an attitude of observing, connecting, and applying reasonable principles to worldly phenomenon.

The *Chemical Professional Laboratory Program*

Fueled by the success of the MATCH program in terms of the positive impact on students and the collaborative efforts between the institutions, UIC invited Harper to participate in another project. Wink and Fetzer Gislason decided their next reform efforts would focus on the general chemistry laboratory program. They had several ideas regarding the focus of the new lab program. However, after a conversation between Wink and Susan Hixson at the New Orleans ACS conference in 1996, during which Hixson indicated one of the proposals was unique, the decision to pursue the Chemical Professionals Laboratory Program

(CPLP) was finalized. The CPLP proposal was funded by the NSF and Wink, Fetzer Gisalson, and Ellefson became co-authors of a new laboratory inquiry program.

As chemical educators with degrees in chemistry, the authors all recognized the significance and relevance of the laboratory component of general chemistry. However, they also knew their students did not necessarily share their innate interest in chemistry. Most students who enroll in general chemistry do not graduate with degrees in chemistry. Instead, they are pursuing careers in the health professions, engineering, biology, or pharmacology. Although the experiments students were completing on both campuses taught the students specific skills and helped them gain hands-on experience with some of the concepts discussed in class, they were quite traditional, did not explicitly promote the process of science, and often seemed irrelevant to students.

The CPLP was designed to provide examples of how chemistry is used by professionals who are not educated and trained as chemists. The idea was to make explicit to the students how chemistry would play a central role in their future careers. The program serves as an example of a collaborative interdisciplinary project that was successful because it was the result of the efforts of a network of faculty in different disciplines and from different institutions. The professional faculty involved in the project were from the departments of biology, chemical engineering, mechanical engineering, pharmacy, medical-surgical nursing, and medical lab sciences at UIC. The chemistry faculty involved in the project met together with all of the "professionals" to share ideas about the lab program and to discuss the specific scenarios associated with the experiment groups. The discussions were stimulating and made the chemists think about some traditional topics in new ways. For example, instead of only performing a typical acid-base titration to standardize a solution, students would use titration to determine the amount of CO_2 produced by the decay of leaves while studying tree leaves and the global carbon cycle. Rather than conduct a calorimetry experiment to verify the specific heat capacity of a metal, students would explore heat capacity as a factor in designing a fireproof safe.

The co-authors shared equally in writing the lab manual that consisted of experiment groups of three labs each; a skill-building lab, a foundation lab, and an application lab. Each author worked individually with different professionals to fully develop the scenario which presents a situation or problem that a non-chemistry professional, such as an ecologist or a nurse, may encounter in their field and to outline the application experiment. As a team, the authors discussed what skill-building and foundation labs would be appropriate for each experiment group. They knew their students were relatively inexperienced in the laboratory, so they designed the skill-building lab to enable students to learn basic techniques, such as preparing serial dilutions or measuring heat capacities that they would utilize in the application lab. The foundation lab was meant to serve as a transitional lab where students would use the technique introduced in the skill-building lab to solve a problem, but still with fairly explicit directions. It is in the application lab that students address the problem outlined in the scenario. The application labs are the least structured. The problem and, in some cases, part or all of the procedure is provided to students, but they are responsible for collecting and interpreting data in order to answer the posed testable question. Additionally, some of the experiments do not have a predetermined outcome. Students are also expected to decide if their data are acceptable and, if not, to make adjustments to their procedure and collect more data. Finally, they are expected to make connections between the scenario, the data they collected, and the chemical principle underlying the experiments.

After a summer of writing, testing, revising, and editing, the experiments were class-tested by collaborators from Oakton College, College of DuPage, and Harold Washington College, as well as UIC and Harper College. The CPLP was also evaluated through direct observations of students and their professors working in the lab and through student focus groups. The students were asked if they were making the connection between their fields and chemistry, if the group work was effective, how the CPLP labs compared with traditional labs and what could be done to improve the program. The study indicated that the connection between chemistry and the other fields from which the experiments are drawn must be made explicit by a discussion of the scenario before starting the experiment group and again at the end of the application lab, otherwise most students failed to draw the connection. However, students did recognize the applicability of working collaboratively with others. Working in a small group enabled them to share ideas about how to collect and analyze data and ultimately enhanced their understanding of both the chemical concepts and the problem posed in the scenario. They also recognized how a lack of preparation by one or more members was detrimental to the group's productivity.

Although the lack of explicit directions in the application lab bothered some students, the improvements they most wanted were associated with how the lab periods were conducted. They wanted more discussion time before and after completing the experiment. For most of the students, this type of lab program was a completely new experience, so they were uncomfortable and lacked confidence with the format; they needed to be more engaged with their instructor and peers in order to develop confidence, to prevent students' feeling like they were left to flounder on their own. Feedback from the students and faculty was invaluable in revising the experiments prior to publication (*6*).

From CPLP to *Working with Chemistry*

The CPLP was published as *Working with Chemistry* (*WWC*), a complete laboratory manual, by W. H. Freeman in 2000 (*7*). Each experiment group, designed to stand alone, was also available separately. An excellent team helped shape the final product. Early adopters suggested one major revision: decrease the number of labs to two per group. The authors, who had learned to listen closely to each other and to other colleagues and students, recognized the validity of the suggestion so in the second edition of the laboratory manual the experiment groups contain only a skill-building and an application lab.

In an effort to further disseminate the essence of the WWC laboratory program, the authors offered workshops sponsored by the Center for Workshops in Chemical Sciences (CWCS). CWCS is another, long standing DUE-funded project, led for many years by Jerry Smith at Georgia State (*8*). Smith approached them to establish a CWCS program in chemical education *per se*, through the workshop "Chemical Education: Supporting Student Laboratory Learning." Workshop participants performed selected WWC experiments and participated in discussions focused on topics such as collaborative learning, inquiry, designing labs, and NSF grant writing. The ultimate goal of the workshops was for the participants to develop a plan for their own innovative laboratory unit or some other laboratory reform. Although participation in the workshops may not have been the sole impetus, some of the participants did submit proposals for and publish papers on innovative laboratory programs and projects.

The "NATS" Initiative

The other projects discussed in this paper were aimed at students taking chemistry as part of a STEM or health-related track. But of course science education is also critical throughout K-12. While NSF has significant work that occurs in research on materials and teacher education for existing K-12 settings, during the 1990's the Division of Undergraduate Education also led the way in getting institutions of higher education to consider how to educate future teachers. The reason for this was based in a simple observation: colleges and universities are the sources of almost all teachers!

NSF's approach included multi-institutional efforts through the *Collaboratives for Excellence in Teacher Preparation.* This funded the *UIC-Community College Collaborative for Excellence in Teacher Preparation* in 1999 under the leadership of UIC's Institute for Math and Science Education. This Collaborative program instituted much more extensive collaboration among UIC STEM and education faculty. In addition, UIC extended its connections with community college partners, including Harper, Oakton, Truman, and Harold Washington Colleges. Several aspects of this effort, including a series of faculty workshops, occurred. Out of this initiative there emerged a group of faculty to address the courses that elementary education majors at UIC take in the natural sciences. However, this work could not just affect courses at UIC, for a large fraction of these majors start at community colleges, including the partners in this grant. So, from the beginning, the "NATS" (short for "Natural Sciences) program was designed by all campuses together.

The NATS courses included three content courses to be taught at one or more of the institutions by STEM faculty. These were named *The Biological World*, *The Physical World*, and *The Chemical World.* In addition, students would take a fourth project-based seminar course that would involve faculty from both education and the natural sciences. In all, elementary education majors would take 14 credit hours of science courses, all within a setting specifically designed for their training as future teachers. The team for the program design was led by Maria Varelas of UIC's college of education. In this case, the collaboration was not just with other departments contributing to chemistry. *All* the disciplines were affected, though in diverse ways.

The actual implementation of these courses was not fully completed during the original grant. But Varelas and the NATS faculty continued the project by securing a CCLI grant, "Integrated Science Courses for Elementary Education and Non-Science Majors." In addition to Varelas, Wink, and Ellefson, the co-PI's on the grant included a faculty member from Harold Washington and from Truman College. But that was just the tip of the iceberg, as faculty from UIC's physics, earth and environmental sciences, and biological sciences fully participated, also. At the different community colleges the courses were taught using traditional settings for general education courses, sometimes using chemistry, physics, or biology designations. But at UIC the courses have been institutionalized as a permanent set. It was possible to get the campus to recognize that smaller sections, additional faculty resources, special student-centered inquiry pedagogy, and special registration procedures were all appropriate for a program serving elementary education majors, since all of this would help those future teachers affect *their* students. The fuller effort has been described in publications (*9, 10*) but here the focus will be on the particular structure and outcomes of *The Chemical World*.

The Chemical World was developed as a general education course that covers many of the traditional topics for a non-majors course through four course units. The faculty recognized early on that, especially for chemistry, the link to elementary education would occur through connections outside of chemistry, since chemistry is not a well-separated part of K-8 education. Specifically, three of the units are on "Chemistry and Life," "Chemistry and Society," and "Chemistry and the Earth." In addition, all of the NATS content courses were built to have an introductory discussion on an aspect of science as a field of inquiry. In the case of *The Chemical World*, this was designated to discuss the question of "Who is involved in science," through a unit involving the "Sociology of Science."

The student focus of *The Chemical World* occurs in two ways. First, the opening discussion of "Who does science?" is built upon a premise that *everyone* is involved in science in diverse ways. In some cases, they use the movie *Lorenzo's Oil* to focus students on two things they may not have thought of before (*11*). On the one hand, the story involves parents who need to learn chemistry and to implement chemistry-based solutions for their son's disease. Through this, they can focus on how students themselves may one day need to know enough chemistry to make difficult decisions about medical care for themselves or for a loved one. On the other hand, the movie also highlights the way many different factors contribute to decisions about what science research is done. As a result, students are alerted to the possibility that, through their taxes, their fundraising efforts, and their advocacy, they themselves are indirectly responsible for deciding what science is done.

The course also includes an individualized project. This is known as the "Big Theme" for the students and is developed in dialog with the faculty over the course of the whole semester., These projects often concerned health-related issues, but students have also chosen to work on environmental issues, questions related to nutrition, and the chemistry of things like how paints are made. The "Big Theme" projects have also enabled students to influence the *content* of the courses. For example, some students have looked at questions of how particular drugs interact with the central nervous system. As a result, more attention is given to the question of drug action via receptor binding. Similarly, different aspects of metabolism are interesting to students, so more attention is focused on why there is different caloric content in different foods, rooted in the course content on molecular structure.

In contrast to the materials development focus of MATCH and *Working with Chemistry,* though, NATS is a more complete curriculum project. Therefore, results have been disseminated mostly through peer-reviewed publications, about the project and findings. Three examples show the breadth of work that can be shared in this way, including: course development; the nature of courses; and chemical education research on the course initiatives impacted students.

The first example concerns the actual development of the courses. The NATS courses have several unique aspects, as noted. But they were not designed *de novo:* their initial design included careful attention to the literature on teacher preparation, and in this case a critical component of their thinking was rooted in a *different* NSF-funded project, at the University of Michigan-Dearborn. Through a careful examination of the work done there by Gail Luera, Charlotte Otto, and their coworkers, they were able to obtain good insight into both the overall multiple-course design and also into the use of projects for student learning (*12*). Thus, their connection with the wider CCLI program occurred as they sought to adapt-and-adopt their work.

The second example concerns sharing what has been done in the courses with other educators. To date, no fewer than four papers have appeared on the nature of these courses, ranging from the general structure to particular examples of how labs are implemented. This, of course, is the core way that CCLI projects can be shared—discussing "How they did it." But in all cases they include particular information on what students do in the courses, another example of their student-focus in action (*9*).

Finally, NATS has produced papers that directly contribute to the chemical education research literature. In one case, a multiple case-study approach was used to describe the breadth of work that fits within the idea of the "big theme" project. This paper used a particular theoretical framework—feminist pedagogy—to frame a description of the project and to carry through with the data collection and analysis (*13*). In another case, a detailed coding of student journals on course topics provided results on the nature of student reflections and on how the frequency and the *type* of reflection correlated with student success. This showed that more frequent reflection correlated with higher grades and with a standard survey-based assessment of metacognition. But, this only occurred if the reflection was related to classroom events or the students' own thinking. Reflection on information in a textbook or other external resource had a negative relation to metacognition and course outcomes (*14*).

At This Point in the Trajectory: Three Projects, Three Biographies, One Community

The three projects discussed here are all, in their own way, continuing to be vibrant parts of the chemical education community. The *Practice of Chemistry* underwent several internal revisions before it was released for mass publication. It continues to be used at many colleges and universities today and is in the process of revision for a second edition.

Working with Chemistry is used by different institutions, primarily as separates. UIC used the experiments as the basis for its general chemistry laboratory program for years. Ellefson adapted experiment groups from WWC for use in liberal arts chemistry courses. The laboratory reform workshops associated with WWC have encouraged other chemical educators to develop innovative laboratory programs to improve student learning and their laboratory experiences.

The NATS program is fully institutionalized, as noted. It has evolved some at the different institutions; as this chapter was being written, Wink was teaching a further version of *The Chemical World* at UIC to another 50+ elementary education majors, including even tighter integration with biology.

Wink started his trajectory with the least training in education and as a research-focused assistant professor. A career shift in 1992 put him in a position to focus on education, but research, funding, and dissemination remained important to his work. His chemical education trajectory, especially through close work with Fetzer Gislason and Ellefson, meant that he also learned to work across departments, across institutions, and across disciplines. He continues to participate fully in NATS and his K-12 work, rooted in that project, now includes extensive work with high school science teachers. In addition, having gained experience in chemical education research, he was in a place where he could join in UIC's Learning Sciences Research Institute, a fully cross-departmental academic program that emphasizes 'Learning in the Disciplines.'

Fetzer Gislason has necessarily changed a lot in these intervening years. In retrospect she sees that her time spent with students has broadened her thinking while her immersion in educational activities has informed her teaching. Each new project has made her more student-focused and open to the ideas of others. She acknowledges a deeper understanding of students and the barriers to their learning and she applies this knowledge in her teaching. Her experience authoring books has made her more cognizant of exact word meanings and better able to express herself succinctly and to insist that students do the same. Most importantly she has come to believe that a solid understanding of math is crucial to student understanding of chemical relationships. Fetzer Gislason is retired from teaching at UIC but remains actively involved with chemical education. She and Wink are working on a second edition of the *Practice of Chemistry*, one with hopes for an extensive electronic upgrade. She is also supervising firstiyear student teachers in the Teach for America program. It is with enthusiasm that she applauds all those schools and personnel who instruct and form their budding science teachers. Fetzer Gislason reports that the student teachers she currently supervises demonstrate attitudes and techniques that have taken the chemical education community years to develop.

The MATCH program was the first NSF-funded project in which Ellefson participated. Since then she has had the privilege of working on several other projects funded by the NSF with Wink and Fetzer Gislason as well as others with whom they networked primarily as a result of that first grant. Her work on these projects has shaped how she interacts with students and colleagues. She has become a much more reflective teacher and one who listens to her students to try to understand where they are, where they want to go, and how she can help them achieve their goals. Ellefson is open to implementing changes in order to become a more effective educator and to improve the learning experience of her students. She encourages the faculty she mentors to also reflect on their teaching, to explore innovative strategies, to observe each other and to engage in discussions about effective teaching. Ellefson learned the tremendous value of collaborating with colleagues and continues to build relationships across disciplines and institutions as she engages in other educational projects primarily focused on continued laboratory reform and assessment.

Finally, the trajectory of these educators over the past twenty years has included significant impact on the chemical education community. Some of this has been national, as Wink and Ellefson are both participants in national organizations for the field. But some of it is also local: the chemical education community of northeastern Illinois now has several examples, in part through the projects described here, that bring together different institutions in both common work and also in discussions about their students. These projects, and others like them, therefore have had, as NSF hopes, a broad impact indeed.

Conclusion: Themes of a Trajectory

The introduction of this paper discussed how the authors have framed their work with four key themes: interdisciplinarity, collaboration, student-focus, and dissemination. In all three examples presented here, these four themes occurred in different ways. Yet, the authors think that, if their twenty-year multi-project experience has taught them (and their many collaborators) anything, it is that there are important general ideas to be seen about how to blend all four themes.

Put simply, the history of these projects supports the following simple assertion: The process of learning by students, which is the only meaningful focus of educational work, is multidimensional, involving different disciplines and blendings of content, pedagogy, and premises. Therefore, strong collaborations are needed, and these have to incorporate different perspectives, different kinds of expertise, and different roles. Yet even the best collaboration is not self-contained in focus. Rather, collaborations must include well-structured plans to make use of the literature and to demonstrate the importance of the project outcomes to the wider community through dissemination.

It is the authors' experience, grounded in the data of their work, publications, and textbooks, that projects that align with this assertion have particularly high potential for broad impacts, with students, the collaborators, and their community. They like to think that this is what the NSF wanted as a way of accomplishing systemic change, and they feel no small amount of gratitude that the Foundation has given them the chance, through multiple projects including others not described here, to carry out such change.

References

1. Division of Undergraduate Science, Engineering, and Mathematics Division, *Undergraduate Course and Curriculum Development Program Announcement and Guidelines*; National Science Foundation: Washington, DC, 1991.
2. Wink, D. J.; Creagan, B. T. *J. Am. Chem. Soc.* **1990**, *112*, 8585–8586.
3. Kwok, T. J.; Wink, D. J. *Organometallics* **1993**, *12*, 1954–1959.
4. Wink, D. J.; Fetzer Gislason, S.; McNicholas, S. D.; Zusman, B. J.; Mebane, R. C. *J. Chem. Educ.* **2000**, *77*, 999–1001.
5. Wink, D. J.; Fetzer Gislason, S.; McNicholas, S. D. *The Practice of Chemistry*; W. H. Freeman: New York, 2002.
6. Ellefson, J. Working with Chemistry: A Laboratory Inquiry Program. In *Making Chemistry Relevant: Strategies for Including All Students in a Learner-Sensitive Classroom Environment*; Basu-Dutt, S., Ed.; Wiley: New York. 2010.
7. Wink, D. J.; Fetzer Gislason, S.; Ellefson Kuehn, J. *Working with Chemistry*, 2nd ed.; W. H. Freeman: New York, 2004.
8. Currently, CWCS continues as *Chemistry Collaborations, Workshops, and Communities of Scholars*. ccwcs.org (accessed February 2013).
9. Varelas, M.; Plotnick, R.; Wink, D.; Fan, Q. *J. Coll. Sci. Teach.* **2008**, *37* (5), 40–47.
10. Plotnick, R. E.; Varelas, M.; Fan, Q. *J. Geosci. Educ.* **2009**, *57* (2), 152–158.
11. Wink, D. J. *J. Chem. Educ.* **2011**, *88*, 1380–1384.
12. Otto, C. A.; Luera, G. R.; Everett, S. A. *J. Sci. Teach. Educ.* **2009**, *20* (6), 537–552.
13. Wink, D. J.; Ellefson, J.; Nishimura, M.; Perry, D.; Wenzel, S.; Choe, J. H. *Feminist Teach.* **2009**, *19*, 31–46.
14. Dianovsky, M. T.; Wink, D. J. *Sci. Educ.* **2012**, *96*, 543–566.

Chapter 9

The Evolution of Calibrated Peer Review™

Arlene A. Russell*

Department of Chemistry and Biochemistry, University of California, Los Angeles, California 90095
***E-mail: russell@chem.ucla.edu**

This chapter reviews the development of the Calibrated Peer Review™ system, which developed out of the NSF-funded Molecular Science Project – one of several Systemic Initiatives for changing the teaching and learning of chemistry. The chapter charts the development of this educational tool for managing student writing and reviewing from the inception of the idea to its implementation in large lecture environments, and its evolution as a resource to broadly facilitate teaching graphical as well as textual communication skills. It also summarizes research into the effectiveness of the method and reasoning about why it is effective as a teaching and learning tool. The overall usage of this system, including in areas outside of STEM is also tracked and presented.

"Calibrated Peer Review™ (CPR) is a web-based, instructional tool that enables frequent writing assignments in any discipline, with any class size, even in large classes with limited instructional resources."

Thus, begins the website introduction (http://cpr.molsci.ucla.edu) to an instructional tool conceived in the early 1990's before Wikipedia (2001), and when MS Office, ChemDraw, and Spartan were being delivered on discs and installed on each computer. Proposing web-based instruction as a goal of the Molecular Science Project in 1995 (*1*) was daring and high risk. Fortunately, the Chemistry Division of the Division of Undergraduate Education under the leadership of Susan Hixson, was willing to take the risk; the ensuing years have validated that choice, as many of the then ambitious goals have become expected teaching practices fifteen years later.

© 2013 American Chemical Society

Inception

The learning goals of the Molecular Science Project were to prepare students (1) who would have a deep understanding of chemistry concepts and principles, (2) who had learned collaboration skills through doing chemistry, (3) who could use the modern technology tools of the chemist, and (4) who could write about chemistry. The radical component of the vision, however, was that the project would meet these goals through the integration of technology and telecommunications into the instructional process, and shift the instruction from lecture to active student learning. Faculty from six institutions (Crossroads School – Joe Wise, East Los Angeles College – Carcy Chan, Pasadena City College – Victoria Bragin, Mt. San Antonio College – Eileen Di Mauro and Iraj Nejad, California State University, Fullerton – Patrick Wegner, and the University of California, Los Angeles – Orville Chapman and Arlene Russell) formed the scientific core of the project. This cross-section represented the common and shared responsibility that still exists for teaching the first two years of chemistry and the diversity of students in the nation. Even by the second year of the project others had come under the Molecular Science umbrella: community college faculty from Albuquerque – Marie Villarba; San Francisco – Tim Su, Houston – John Magner, Seattle – Joann Romascan, and Las Vegas – Carolyn Collins were active participants, through an "Adopt and Adapt" FLASH grant in 1999 (*2*), which formalized the collaboration; Mike Mosher at the University of Nebraska, Kearney soon began to contribute to the project also through the "Adopt and Adapt" program at NSF (*3*). While the community college faculty saw the CSU-Fullerton on-line Mastering Chemistry homework and active learning products as the primary boons for their students, Mosher seized the centerpiece project developed at UCLA, Calibrated Peer Review™, to teach students scientific report writing. He utilized the Calibrated Peer Review program to manage the instruction and writing in an innovative, intensive-writing laboratory curriculum. For each experiment, students focused on a different component of a lab report, thus providing manageable, scaffolded instruction throughout the term leading to the proficiency expected in upper division courses. The dilemma Mosher was facing then on how to teach technical writing to large classes, and to give students practice writing and targeted feedback, is still shared by many faculty. Scientific report writing forms the core of writing in lower division chemistry courses, yet these are the courses that carry the highest enrollments. Mosher's foresight, to use CPR to teach report writing skills, has been replicated in other institutions and disciplines in the ensuing years. A review of the assignments created in the original CPR program shows dozens of assignments connected to laboratory experiments and to the communication of scientific data. A few of the CPR users have published the results of their work (*4*, *5*).

CPR, however, is not restricted to implementation in existing writing arenas. The topics of assignments are far ranging, and the use of CPR is limited only by the creativity of the faculty who have envisioned writing as a way of learning (*6*, *7*). Not only have more than 300,000 students used the program in over 1500 institutions around the world, many instructors have used CPR as a research tool to investigate and improve student learning.

Calibrated Peer Review, which is based on the scientific practice of expert peer review, is the most enduring of the products of the Molecular Science project. Earlier articles have elaborated on the details of the CPR process (*8–12*). We provide here a brief synopsis of the process to give context to the evolution of the program and the research in learning that the tool has facilitated. Although the program has evolved in response to users' needs and changing technology, the embedded pedagogical structure has proved sound.

The Calibrated Peer Review Process

Writing-across-the-curriculum and adjunct-writing courses continue to have strong and widespread advocacy in higher education (*13*, *14*). These courses are, however, resource intensive and writing in the STEM disciplines has often been relegated to the upper division or the graduate level, particularly in large institutions. Through web-based management of the text submission and review process, CPR was the first program to enable writing in any size class without additional teaching resources. The success of a CPR assignment relies on precise and clear articulation of the topic of the assignment. The aphorism "clear writing demonstrates clear thinking" captures the pedagogy of CPR, that writing provides a window on student understanding of the topic and an alternative means to assess learning. To do this, the CPR program is grounded on the precept that peer review is first and foremost a fundamental instructional strategy for engaging students through critical thinking *and* secondarily a mechanism for evaluation of their peer's writing. Effective peer review, however, depends on "qualified judges" (*15*). Training students to become those qualified judges or capable reviewers, that is "CALIBRATING" them to understand the nuances of the content of the assignment and to recognize errors and misconceptions, *when they have just learned about the topic,* constitutes the signature component for each and every writing assignment. It is fundamental to the pedagogy. Monitoring that training, providing feedback on the training, and tracking the training so that poorly trained reviewers carry little weight on the reviews they give their peers give confidence to the reviewers and reliability to the reviews.

As an instructional tool, Calibrated Peer Review serves to teach higher-order thinking skills through scaffolding the writing and evaluation tasks. An assignment consists of four components that are interwoven to encourage learning at each stage (Figure 1). Students are first presented with a writing task, which is supported by resources, references, and guidance for preparing for the actual text writing. This guidance enables students to address the topic of the assignment, yet gives them freedom to articulate their ideas in their own words. Even students recognize the power of writing-to-learn (*12*), which undergirds the strong and widespread advocacy for writing in higher education (*16*, *17*), and serves as the initial motivation for many faculty to try using CPR.

The second stage of the CPR process, the Calibration training, begins only after students have wrestled with the writing and submitted their work for review. The program presents the students with three pre-written texts, which span the range of responses expected from their peers. Studying and evaluating these texts not only prepares them for peer reviewing, it also provides a directed opportunity for students to continue to learn the topic. Generally one text will be an exemplar, and the other two will contain the common errors and misconceptions that frequently occur. Students evaluate and rate these texts using a rubric that addresses the critical issues and concepts of the topic. Prior guidance on how to rate texts helps to provide the common understanding that experts use to value student work. The consistent goal of the training is to bring all students to an "expert" level of understanding. Realizing that the rate of deep learning is not necessarily correlated to mastery, the CPR program does not penalize students who require a second attempt at mastering the training component of the program. Even with a second attempt, not all students will make "expert status." Thus, behind the scenes, the program tracks student calibration performance and assigns a "reviewer competency index" (RCI) to each student. The RCI subsequently is used as a weighting factor in the scoring algorithm to determine what impact the reviewer will have on the peer's text rating.

The third stage of the CPR process involves the double-blind review of three peer texts. The program randomly selects from the full set of texts only after all texts have been submitted. Laggards in text submission are just as likely to be assigned a text from an early bird as from a text submitted any time before the due date. Students then use the peer assessment rubric, which they learned during calibration training. In this stage, however, students are required to give written feedback to their peers on the reasons for their assessments and ratings. As in the academic world, the most useful reviewer's comments can serve to support improvement. With CPR, good feedback from a peer's voice can be a powerful learning object. For example, a student in an organic chemistry class wrote:

> **Rating:** 4; **Explanation:** You have the general concepts of base peak and molecular ion peak down, but your explainations [sic] of how you obtain these peaks is incorrect. First, when you determine the molecular ion peak, you add the weights of the ions that were given with the problem (i.e. C=12, not 12.01, H=1, not 1.01, etc.). Hence, your molecular ion's weight will be 84, not 84.93. Although your concept is correct, your calculation is wrong. Also, your explaination of the 86 m/z and 88 m/z readings is very difficult to follow. You never directly tie the 86 m/z reading to CH2Cl35Cl37+. Your explaination of the differing levels of 86 m/z and 88 m/z is good in that you correctly state that it has to do with the ratios that the isotopes occur in nature. The details of your explaination, however, are incorrect. The fact that there are two Cl37 atoms, which have a 25% chance of occurring in nature, present in the molecule LOWERS the chance of that molecule being present, since there are two "one in four" chances that have to be overcome instead of just one in the case of only one Cl37 atom being present. You do not add the 25% chances together and get a 50% chance. Also, you have a few spelling errors and you neglect to put the "plus radical (+.) after each ion. You also don't have a summary sentence. Additionally, you lack an explaination of why low pressure upheld in the machine (it is done to

reduce the amount of intermolecular processes). Overall, your general understanding is good, but you need to work on the details.

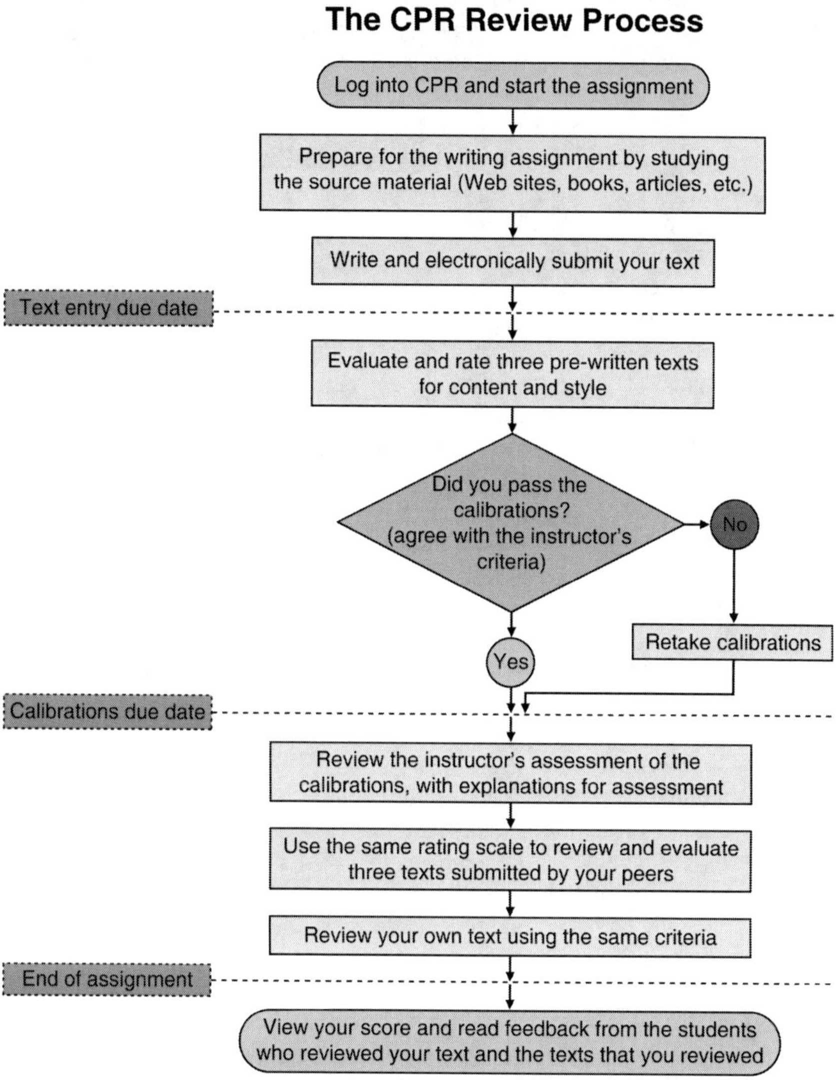

Figure 1. Tutorial flow-chart of the process and stages of an assignment as seen by a student.

By utilizing the role of feedback to guide revision of written work, some faculty have developed assignments for lab reports where the CPR process constitutes the "pre-write" and drafting stage of the report. Only the final report, with responses to the peer feedback, are submitted to the instructor for grading (*18*). The next version of CPR, which is in development, will include a revision feature as one of the options, which faculty can select, for an assignment.

The final stage of the CPR process brings closure to the assignment for the student. After training and reviewing peers' texts, each student has the opportunity to critically evaluate their original work in light of their new understanding. The articulation of conceptual change by recognizing problems in their original ideas solidifies the progress towards a greater understanding.

CPR4 and CPR5: Augmenting and Enhancing Calibrated Peer Review

In 2004, a major restructuring of the delivery mechanism for CPR began. It had become apparent that many institutions were interpreting federal regulations (FERPA) (*19*) as a restriction on student work and grades from existing outside of the campus technology firewalls. Because all CPR data at that time were stored at UCLA on a master server, faculty were prevented from using the program. In particular, the faculty at Texas A&M, which had endorsed CPR as a mechanism to meet their new academic senate requirement for writing in every discipline, required a new approach. As a short-term solution the University of California and Texas A&M entered into an agreement, which allowed the latter to house and use a copy of the program on their College Station campus. With one problem solved, another arose. The Texas A&M faculty were now able to use the program, but were isolated from the shared community of users who were developing and modifying discipline-based assignments. Likewise, others could not benefit from their intellectual creativity in developing assignments as part of a science and math initiative (*20*).

The concept of a two-server distributed version (CPR Central and CPR Local) of the program emerged. Once again NSF responded and supported the vision of a shared community of users (*21*). CPR Central, located at UCLA, was developed to provide a place for authoring, storing, and sharing assignments. This central assignment library includes all assignments that were part of the old CPR server library as well as a place to continually grow assignments that are created by the community of CPR authors. Figure 2 contrasts the assignment resources now available through the Central Library with the number of assignments that have been shared in the original CPR program over 15 years.

CPR Local became the entity of the program that was designed to be installed on a server at the user's institution. Students' records and work are now stored entirely on the host campus, safely behind its firewalls. The institution copy of CPR Local communicates with CPR Central only when an instructor is setting up an assignment for subsequent class use. After an assignment is copied to the CPR Local server, the institution is no longer dependent on a server at UCLA and students no longer have to share the UCLA server's processing power and resources with other institutions.

The revision and rewrite of the CPR program furnished an opportune moment to add many new features for students, for instructors, and for assignment authors. Perhaps the most important new feature for students in CPR4 has been a new function at the end of an assignment for students to see how others reviewed the texts that they had pondered over and evaluated. Evaluation is new and often intimidating to students. The hope that students would gain confidence in their own evaluation skills when they saw that others detect the same strengths and weaknesses in their peers' work as they did has been borne out. In a 2008 pilot test of the tool, 65% of the students agreed or strongly agreed that their "confidence increased by comparing others' reviews." They value this new feature for a variety of reasons (Figure 3). Students are relieved to find that their peers provide reviews consistent with their own.

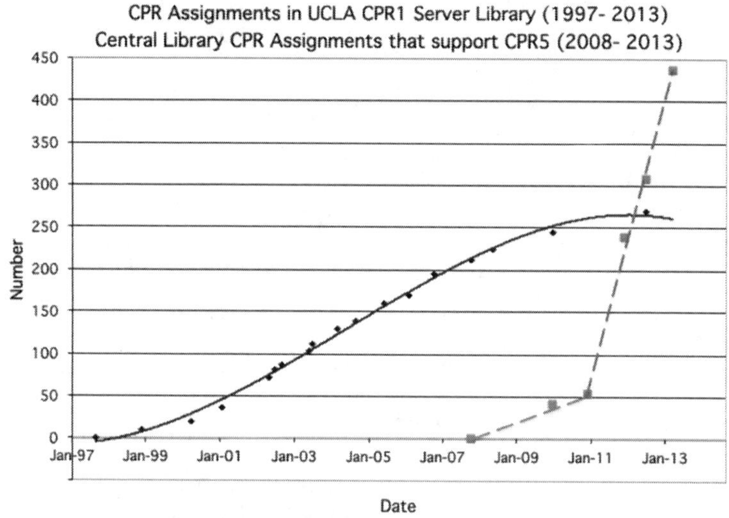

Figure 2. Comparison of the growth and number of assignments available to the community of CPR users in the original CPR program (solid line) hosted at UCLA and in the shared Central CPR Library (dashed line) used with the distributed CPR4 and CPR5 versions of the program.

The most innovative feature of the new distributed CPR4, however, was a new tool that addressed a faculty need. The Texas A&M collaboration had shown that faculty scholarship was an integral part of CPR assignment development (*22*). Assignment authors bring to the process a deep knowledge of content and a recognition of student learning. Because of the need to know how students struggle with the topic, well-crafted assignments rely on the wisdom of experienced instructors. As the faculty brought to bear their creativity as authors they wanted a way to document this new scholarship of teaching (*23*). They also asked how to give credit to the authors whose assignments they were adapting or adopting. Plagiarism of others' intellectual ideas is not acceptable.

Because the hours, effort, and teaching experience necessary to create a successful assignment, as well as the need for fair attribution of scholarly work requires recognition, a citation index function was built into the new CPR Central Library. New and old users can now search the database of hundreds of assignments and ethically use, copy or modify an existing assignment. The program automatically maintains a record of the usage of each assignment by the original author, and the derivative works that have emanated from the original creation. As more schools require accountability for teaching in tenure and promotion portfolios CPR can be used to document the impact of an assignment beyond an author's campus.

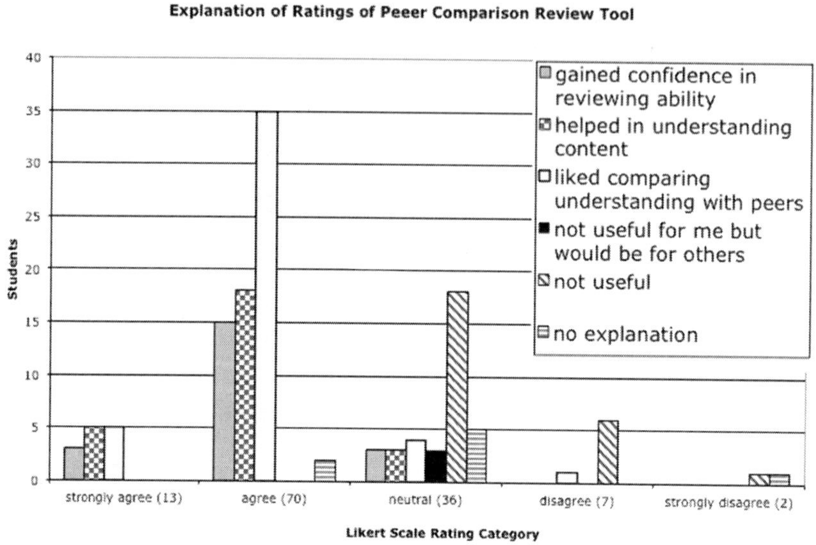

Figure 3. Student explanations of the reasons on the Likert scale ratings of the statement, "Comparing my peers' ratings of the texts I reviewed improved my confidence in my understanding" [of the topic of the assignment]. (gray – "gained confidence in my reviewing ability; checkered – helped in understanding content; white – liked comparing understanding with peers; black – not useful for me but would be for others; diagonal – not useful; horizontal – no explanation).

Development of Calibrated Peer Review, Version 5 followed logically and quickly after CPR4. The rapid growth in networking technology, in both software and hardware, had finally enabled the feasibility of creating a tool to support an endemic component of scientific writing—visual representations of graphs, tables, pictures, spectra or other images. However, the new tool that allows students to upload a file with their text submission or in lieu of text opened the door for other uses for CPR. The program is no longer limited to writing and writing assessment. Driven by the need for Engineering departments to assess multiple forms of communication, CPR5 became a vehicle to handle such a process (*24*). In CPR5 file uploads can have any format; what students upload to their campus CPR server is limited only by the requirements of an assignment and local policies. The program is robust. Files with more than 11 different extensions were successfully used by students during the first pilot test of this new feature. Thus, the evolution of CPR in response to STEM needs, has broadened its applicability to peer evaluation of posters, PowerPoint slide decks, videos, oral presentations, and music. New and creative uses continue to appear. The future promises tools that more fully adopt the revision processes of scientific publications.

Research on Learning Using CPR

Creation of new CPR tools and materials has not occurred in a vacuum. Since the inception of the program, faculty have been concerned with the effectiveness of a web-managed, peer-review process. As well as instructors, institutions are increasingly being held accountable for assessment of student learning. They see CPR as a way to show they are meeting their learning objectives. Again Engineering is on the forefront. For example, CPR documentation is being used in ABET accreditation reviews at UCLA and in at least one other Engineering department (*25*).

Much of the research on the impact on learning using CPR has focused on test scores on the topics of the CPR assignments (*26–28*). The studies repeatedly show exam score increases of the order of 10%. Pelaez instituted a "time series design" for a single class of 42 students. She alternated topics taught using didactic lectures and CPR assignments with lectures, group work and multiple choice quizzes to provide the assessment feedback intrinsic in CPR. Pelaez found that students performed better on exams on the topics of the course that included CPR assignments than on the questions on topics where the lectures were augmented with group discussion and quizzes. She saw similar gains across all levels of student ability on the multiple choice questions on exams, but the top performing students (on exams) had larger gains on the essay questions than the weaker performing students (*26*). Chapman's early work in Economics gave the first insight into the additive learning effected by the evaluation components of CPR. His "intact class comparison" involved three large classes (>100) taught by the same instructor using the same ten case studies in all classes. One of the classes was assigned the CPR writing component only; the other two classes completed full CPR assignments with peer evaluations and self-assessment. Chapman repeated the experiment the following semester. He observed that on common exams, across every quintile, the grades for his students were higher in those sections that had completed the full CPR assignments rather than just writing about the topic before the class discussion (Figure 4). Rudd *et al* carried out a similar study in an introductory Geology course for non-science majors with two sections taught by the same instructor. Like Pelaez and Chapman, they found that scores on essay questions on exams were significantly higher for the group who had completed full CPR assignments than the group who had only written about the topic. Performance on multiple choice questions was also higher for the CPR group, but the difference was not statistically significant (*27*).

Others have also documented the impact of the peer review process on learning. When students' opinions on their learning gains from using CPR are collected, the importance of peer review always surfaces:

> *"Truthfully, I enjoyed CPR. Writing the essay was a great way to review the specifics of cyclohexane strain, but when I reviewed the peer essays I think I learned most. I was able to see what strong points my essay had, and more importantly where I was lacking (29)."*
>
> *"When I read other people's essays, I see 'Oh, this is what tied into that idea' like the streamlined body makes [penguins, et. al.] move faster. If I*

didn't mention that [point] in my essay or didn't realize that, then reading someone else's essay is what showed me that that's how I was supposed to answer the question (30)."

More recently Enders et al. (*31*) carried out a detailed analysis of learning throughout an assignment. They found that the top students in their graduate statistics class learned the content of the assignment (linear regression) during the calibration training stage, while the median and low-performing students did not master the concepts until the self-assessment phase of the assignment. Their work parallels the findings of this author in a 2010 study in which students assessed the quality of Beer's Law graphs prepared by their peers (Figure 5).

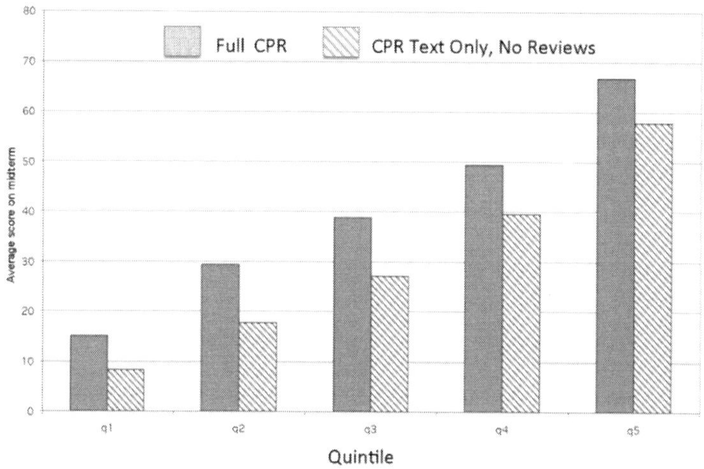

Figure 4. Comparison of midterm performance by students who only performed the writing component of CPR assignments and those who completed the reviewing stages also. Graphing the scores by quintile show that the gains occur at all levels. Economics 200, Fall 1999. K.S. Chapman, COBAE Faculty Report (1999-2000).

Seventy-one percent of the students whose original graphs did not meet expected standards, learned through the calibration and reviewing stages of the assignment and subsequently identified the weaknesses and problems in their own work (*33*).

Writing-to-learn and learning-to-write are not synonymous. While "clear writing demonstrates clear thinking," clear writing can also demonstrate erroneous thinking and poor writing can obfuscate correctly understood concepts. Improving students' science writing skills has been a driving force for adoption of the CPR process for many instructors. Most reported studies have found that writing clarity does improve when courses have used CPR (*4–6*, *33*). However, the results of studies that have teased out writing-to-learn from learning-to-write are mixed. Hartburg, et al. (*5*) found significant writing gains for students whose biochemistry reports were evaluated by peers and losses for those whose reports were graded by teaching assistants. Walvoord et al (*34*) found no gains. Reynolds and Moskovitz' analysis of the emphasis or attention placed on writing skills in a random sample of the STEM assignments posted on the CPR website sheds light on these contrasting results (*35*). They found that 90% of the assignments lacked clear expectations for writing in the prompt students were given. Rather, the writing prompts and evaluation focused on content. Although no guidance was provided in the prompt, some assignments had implicit writing expectations. However, even those assignments that did address writing quality in the rubric tended to concentrate on the lower-order skills of mechanical and grammatical errors. They found that few assignments addressed the higher-order writing skills of effective argumentation, the use of evidence-based analyses, organization, or appropriate use of sources. Attention to audience also seldom occurred.

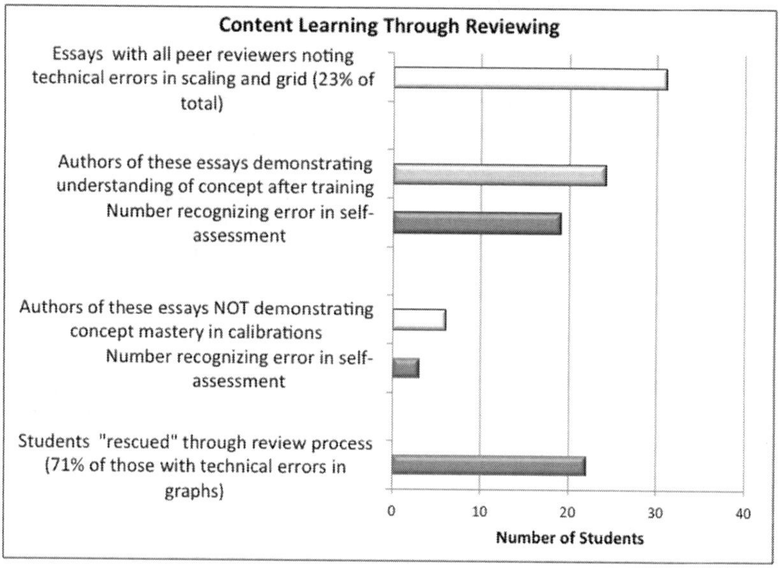

Figure 5. Location in a CPR assignment (white – text entry, light gray – calibration stage, dark gray – post review) where students demonstrate understanding of graphing skills. Twenty-four of the 31 students showed understanding at the end of the calibration training stage although not all (only 19) recognized the error in their own work. Three additional students learned the principles when reviewing their peers' work and correctly identified the errors in their own work.

That students' higher-order writing skills do not improve when they are not explicitly addressed should not be surprising (*36*). Instructors of large lower division classes know well that students generally do only what they are asked to do. Therefore, explicitly articulating to students the writing expectations and providing them with feedback on their writing performance is just as important if an instructor's goals for a CPR assignment include scientific writing skills. In an on-going study at UCLA, higher-order writing skills are being imbedded into both the guidance and the rubric of CPR assignments. This first step has established that 100% of the students believe their own essays "have logical flow." Their peers often disagree! However, when the writing guidance specifies an explicit trait that leads to logical flow, such as the use of transitions, the idea of logical flow becomes tangible and students are able to recognize this quality or lack of it in their own writing. Student use of transitions and other elements that improved organization increased and their self-assessment showed no significant differences with the assessments from their peers' evaluation of their "logical flow." Students will attend to writing if asked to do so and can improve their writing skills if the practices inherent in scientific writing are properly scaffolded into the instruction.

Finally as the knowledge in the STEM fields expands exponentially, teaching students to have confidence in their ability to be independent life-long learners becomes more critical. Likkel recently reported on the impact of using CPR on students' confidence in their abilities (*37*). Like Chapman, her study involved three intact classes. Two used CPR for their writing assignments, for which she provided the feedback. Her other two astronomy classes used CPR to manage and assess the four essays in the course. Table I shows her students' changes in their confidence in knowing whether they had written a "good" essay and in their perceived skill in evaluating their own work. On both criteria, those who had the opportunity to engage in active evaluation and peer review were more positive about their ability to assess their own work when they left the course.

References to other studies using the Calibrated Peer Review program as a research tool may be found on the program website (http://cpr.molsci.ucla.edu).

Table I. Comparison of the change in confidence in one's ability to recognize the quality of one's own writing and in one's knowledge of the skills necessary to assess one's writing by students who received instructor feedback on their four writing assignments (non-CPR) and those who reviewed peers' texts and provided feedback to their peers. L. Likkel J. Coll. Sci Teaching, 41, no. 3 (2012)

*Perceived ability to tell if own essay good**			*Perceived skill in assessing own writing**		
	CPR (54)	*Non-CPR (21)*		*CPR (88)*	*Non-CPR (30)*
more positive	65%	29%	more positive	43%	23%
no change	20%	38%	no change	51%	77%
more negative	15%	33%	more negative	6%	0

* Excludes students who indicated they were 'very confident' at both start and end of term.

Conclusions

During the past 15 years, student writing and peer review using the CPR program has been employed by more than 300,000 students in over 1500 institutions and 3500 courses. What began as a convenience for economies of scale has serendipitously evolved for many instructors as a more powerful method to teaching critical thinking within their disciplines. It has been used as a research tool for assessing learning, for improving instruction, for documenting adoption of innovation, and most recently for validating new digital lexical analysis tools. The program has found application beyond the STEM disciplines (Table II) and is expanding into other areas of communication. CPR Version 6, which will allow for revisions after review portends to bring this instructional resource even closer to the authentic practice of science. Lichter asserted that "the goal of peer review is improved products (*15*)." The goal of CPR remains steadfast: to increase students' learning and their understanding of the importance and validity of peer review in the improvement of scientific knowledge.

Table II. Discipline distribution of the 500 assignments in the CPR Central Library, March 2013

Astronomy	5	Computer Science	1	General Science	20
Biochemistry	4	Earth and Space Sciences	7	NON-STEM	
Bioengineering	2	Engineering	60	Social Sciences	58
Biology	91	Environmental Science	7	Humanities	47
Physiology	7	Mathematics and Statistics	31	Medicine, Dentistry, Public Health	44
Chemistry	115	Physics	13	Other	9

References

1. Molecular Science Project. National Science Foundation, DUE 95-55605.
2. Bringing Systemic Change to Community College Chemistry. National Science Foundation, DUE 99-50320.
3. Implementation of a Writing-Intensive Chemistry Laboratory Curriculum. National Science Foundation, DUE 99-50674.
4. Furman, B.; Robinson, W. Improving Engineering Report Writing with Calibrated Peer Review™. In *Proceedings of the 33rd Annual Frontiers in Education Conference*; Budny, D., Ed.; IEEE Digital Library: Piscataway, NJ, 2003; pp F3E14–F3E16.
5. Hartberg, Y.; Guernsel, A. B.; Simpson, N. J.; Balaster, V. Development of student writing in biochemistry using calibrated peer review. *J. Scholarship Teach. Learn.* **2008**, *8*, 1, 29–44.
6. McCarty, T.; Parkes, M. V.; Anderson, T. T.; Mines, J; Skipper, B. J.; Grebosky, J. Improved patient notes from medical students during web-based teaching using faculty-calibrated peer review and self-assessment. *Acad. Med.* **2005**, *80* (10) (suppl.), S67–70.

7. Clase, K. L.; Grundlach, E.; Pelaez, N. J. Calibrated peer review for computer-assisted learning for biological research competencies. *Biochem. Mol. Biol. Educ.* **2010**, *38* (5), 290–295.
8. Robinson, R. An application to increase student reading and writing skills. *Am. Biol. Teach.* **2001**, *63*, 474–480.
9. Russell, J.; Pearson, M. Instructional technology jewels. *J. Coll. Sci. Teach.* **2004**, *33* (7), 24–28.
10. Russell, A. A. What Works - A Pedagogy: Calibrated Peer Review. *Project Kaleidoscope*, 2004. http://www.pkal.org/documents/Vol4CalibratedPeerReview.cfm (accessed May 24, 2013).
11. Russell, A. A. Calibrated Peer Review™: A Writing and Critical-Thinking Instructional Tool. In *Invention and Impact: Building Excellence in Undergraduate Science, Technology, Engineering and Mathematics (STEM) Education*; American Association for the Advancement of Science: Washington, DC, 2005. http://www.aaas.org/publications/books_reports/CCLI/ (accessed May 24, 2013).
12. Russell, A. A. Reinvention Center Spotlight: Creative Uses of Instructional Technology. *Calibrated Peer Review™: Honing Students Critical Thinking Skills*, 2011. http://www.sarc.miami.edu/ReinventionCenter/Public/Spotlights/?e=41 (accessed May 24, 2013).
13. Barnett, R. W.; Blumner, J. S., Eds.; *Writing Centers and Writing across the Curriculum Programs*; Greenwood Press: Westport, CT, 1999.
14. Berrett, D. An Old-School Notion: Writing Required. *Chron. Higher Educ.* **2012**, October 12.
15. Lichter, R. *Symposium on Peer Review, 220th National ACS Meeting*, Anaheim, CA, March 22, 1999.
16. Thaiss, C.; Porter, T. The State of WAC/WID in 2010: Methods and results of the U.S. survey of the international WAS/WID mapping project. *Coll. Composition Commun.* **2010**, *61.3*, 534–70.
17. *What Is Writing in the Disciplines?* The WAC Clearinghouse. http://wac.colstate.edu/intro/pop2e.cfm (accessed May 24, 2013).
18. Timmerman, B. E.; Strickland, D. *Faculty Should Consider Peer Review as a Means for Improving Students' Scientific Reasoning Skills*; University of South Carolina, Scholar Commons Faculty Publications; Columbia, SC, March 15, 2009.
19. *Family Education Rights and Protection Act (FERPA)*. www.ed.gov/policy/gen/guid/fpco/ferpa/index.html.
20. *Writing for Assessment and Learning in the Natural and Mathematical Sciences (WALS)*; NSF 02-43209.
21. *Augmenting CPR*; NSF 04-42828.
22. Simpson, N.; Russell, A. A.; Gunersel, A. B.; Layne, J. Calibrated Peer Review™ in math and science: Research and practice. In *Proceedings of the National STEM Assessment Conference*, 2006; pp 266–273.
23. Boyer, E. L. *Scholarship Reconsidered: Priorities of the Professoriate The Carnegie Foundation for the Advancement of Teaching*, 1990.

24. *Collaborative Research: Improving Engineering Communication Education with Enhanced Calibrated Peer Review*; NSF 08-16660, 08-16859, 08-17515.
25. Berry, F. C.; Carlson, P. A.; Millard, D. L. In *Proceedings of the National STEM Assessment Conference*, Calibrated Peer Review for ABET Assessment, 2006, pp 190–193.
26. Pelaez, N. Problem-Based Writing with Peer Review Improves Academic Performance in Physiology. *Adv. Physiol. Educ.* **2002**, *26*, 174–184.
27. Rudd J. A.; Wang, V. Z.; Cervato, C.; Ridky, R. W. Calibrated Peer Review Assignments for the Earth Sciences. *J. Geosci. Educ.* **2009**, *57*, 328–334.
28. Chapman, K. S. COBAE Faculty Report, California State University, Northridge, personal communication, 1999–2000.
29. Organic Chemistry, University of California, Riverside, personal communication, 2000.
30. Rudd, J. A., II; Thomas, K. J. *Integrating Writing Assignments into Undergraduate Science Courses Using Calibrated Peer Review*, 17th BCCE: Bellingham, WA, July 31, 2002.
31. Enders, F. B.; Jenkins, S.; Hoverman, V. Calibrated Peer Review for interpreting linear regression parameters: Results from a graduate course. *J. Stat. Educ.* **2010**, *18* (2), 1–27.
32. Carlson, P. A.; Russell, A. A.; Waggenspack, W. N.; Wilmot, C. G.; Bowles, D. F.; Voltmer, D. R.; Monroe, W. T.; Hull, W. R.; Raubenheimer, C. D. Improving Engineering Education with Enhanced Calibrated Peer Review: Assessment of a Collaborative Research Project. In *Proceedings of the 2012 Annual Conference of the American Society of Engineering Education*, San Antonio, TX, 2012.
33. Prichard, J. R. Writing to learn: An evaluation of the Calibrated Peer Review™ program in two neuroscience courses. *J. Undergrad. Neurosci. Educ.* **2005**, *4* (1), A34–A39.
34. Walvoord, M. E.; Hoefnagels, M. H.; Gaffin, D. D.; Chumchal, M. M.; Long, D. A. An analysis of Calibrated Peer Review (CPR) in a science lecture classroom. *J. Coll. Sci. Teach.* **2008**, *37*, 66–73.
35. Reynolds, J. A.; Moskovitz, C. Calibrated Peer Review™ assignments in science courses: Are they designed to promote critical thinking and writing skills? *J. Coll. Sci. Teach.* **2008**, *38* (2), 60–66.
36. Perry, W. G., Jr. *Forms of Intellectual and Ethical Development in the College Years: A Scheme*; Holt, Rinehart, and Winston: New York, 1970.
37. Likkel, L. Calibrated Peer Review™ essays increase student confidence in assessing their own writing. *J. Coll. Sci. Teach.* **2012**, *41* (3), 42–47.

Chapter 10

A Chronology of Assessment in Chemistry Education

Stacey Lowery Bretz*

Department of Chemistry & Biochemistry, Miami University, Oxford, Ohio 45056
***E-mail: bretzsl@miamioh.edu**

Chemists have published data about their assessment practices since the early 1920s. What to assess and how to assess remained largely unchanged for more than 50 years. As chemistry education research emerged as a distinct subdiscipline of chemistry, assessment provided intriguing data regarding errors in student thinking and launched new lines of research into measuring student thinking. In the last few decades, discipline-specific tools have been developed to measure both cognitive and affective learning in chemistry. Ensuring the reliability and validity of data remain important benchmarks in the discipline. A chronology of assessment in chemistry education over the last 90 years (1924–2013), including its role in chemistry education reform and challenges for the future, is discussed.

The "Irreducible Minimum": 1924−1970

Teachers of chemistry have long been interested in not just sharing what they know, but measuring what their students know, too. In 1924, the very first issue of the *Journal of Chemical Education* published a paper entitled "What We Teach our Freshmen in Chemistry" (*1*) that catalogued the content being taught and tested on final examinations for more than 18,000 students enrolled in college chemistry at 27 institutions. The authors' analysis of final exam questions categorized what they refer to as "the irreducible minimum of chemical knowledge a student must possess to pass the course" (p. 9). More than 1800 final exam questions were

© 2013 American Chemical Society

distributed across four categories as follows: equations and problems (36.2%), descriptive chemistry (26.3%), theory (23.5%), and useful applications (13.8%).

Later that same year, S.R. Powers published an article in which he reported the results of creating and administering what he coined the "Test for General Chemistry" (*2*). Two forms of the test were created, with items drawn from a collection of 350 distinct tasks designed to measure the "ability of high school children to do...chemistry." Powers described the test as:

> "Each form consists of two parts. Part I...composed of 30 items...test[s] a wide range of knowledge including biography, chemical properties, chemical composition, commercial processes, and terminology. Part II consists of 37 items and tests the students' ability to write formulas and equations, to give the chemical name of common substances, to give chemical composition, and to do simple calculations." (p. 139)

Powers reported that all the items could be scored "entirely objectively," and that the items were arranged from least difficult to most difficult, with the difficulty of part I and part II being comparable. He calculated the reliability by comparing student performance on the two forms, and also explored the relationship between grades given by high school teachers and performance on this test.

The fact that these reports can be found in literally the earliest published accounts of chemistry education suggests that assessment of student knowledge has long been a priority for chemistry teachers. Indeed, what is today known as the ACS Exams Institute released its first test (in general chemistry) in 1934 (*3*). In many ways, the creation of these assessment tools permitted chemistry faculty to reach consensus regarding what aspects of student learning they wanted to measure. Powers' report (*2*) also points to the early importance of reliability and accounting for error to chemists in these measurements.

Misconceptions and "The Grim Silence of Facts": The 1970s and 1980s

No additional manuscripts were published in the *Journal of Chemical Education* focusing on assessment for the next half-century. Not surprisingly, the focus on teaching and testing student knowledge of theory, equations, and problems remained intact during this time, especially in the context of post-Sputnik curriculum reform efforts such as the Chemical Bond Approach (*4*) and ChemStudy (*5*). Then in 1970, Derek Davenport issued a call-to-arms in his article entitled "The Grim Silence of Facts" (*6*). Students who had completed undergraduate degrees in chemistry and entered a graduate program in chemistry thought of silver chloride as a pale, green gas. The teaching and testing of descriptive chemistry, i.e., facts, was falling by the wayside in the wake of an ever-increasing emphasis on theory, Davenport cautioned.

Chemists began to search for explanations to this seeming contradiction. How could students be assessed in class after class, even to the point of earning degrees and entering graduate programs, yet still harbor such fundamentally incorrect ideas? The work of J. Dudley Herron to introduce Piaget's theory of learning to chemists stands as a significant event in the history of assessment in chemistry education research (*7*), moving chemistry education reform beyond the facts vs. theory debate:

> "…in my judgment, a large part of chemistry *is* abstractions. The temptation to return to a course based on the blind memorization of a catalog of descriptive chemical facts is as repugnant to me as the continuation of courses based on the blind memorization of inscrutable theory. The alternative…is to recognize why the theory is inscrutable, i.e., a large portion of our students operate [without the ability to carry out abstractions]." (emphasis original, p. 149)

Herron argued that decisions about what to assess, and therefore what to teach, ought to be guided by knowledge of how students learn and theory that describes their development as learners. Students are not to learn just chemistry, but to also learn why they need to know chemistry (*8*).

In the late 1970s and early 1980s, accounts of students' chemistry-specific misconceptions and alternative frameworks (*9*) began to appear more frequently in the literature. Linke and Venz reported fundamental errors in student thinking about atomic and molecular structure, phase changes, and solubility (*10*), as well as light, heat, and chemical reactions (*11*). Osborne and Cosgrove (*12*) described their assessment of students' understanding of phase changes in water. Peterson, Treagust, and Garnett (*13*) reported on students' ideas about covalent bonding and structure. Shortly thereafter, Treagust (*14*) authored his classic, oft-cited paper describing in detail one method for developing and using diagnostic tests to assess students' misconceptions in science.

Treagust's paper marked a clear shift in thinking about assessment. No longer could chemists write tests focused solely on the facts of descriptive chemistry and/or the principles of atomic and molecular theory. Chemists like Davenport, Herron, and Treagust were sounding the alarm that students were able to perform well on these traditional tests and still harbor alarming misconceptions about the chemical and physical properties of molecules. Assessment had broadened from a singular focus upon measuring what students knew to shining light upon what they had learned incorrectly.

As the 1980s came to a close, efforts to reform chemistry teaching and learning could no longer afford to focus on just the 'what' of chemistry. The central challenge was no longer only a debate regarding the relative merits of descriptive facts and theoretical principles. Chemistry education reform now faced overwhelming evidence that students' prior knowledge and current instructional practices were interacting in unintended combinations to yield persistent misconceptions. Chemists had begun to realize that curriculum and pedagogy reform would require thinking more broadly about learning and creating assessment tools that could measure multiple dimensions of learning chemistry. Assessment and the data it provided would be essential going forward.

Concept Learning, Algorithms, and the Particulate Nature of Matter: The Early 1990s

Alex Johnstone (*15*) moved the conversation on assessment and students' understanding of chemistry forward when he proposed a triangle to represent the knowledge of chemistry, with each corner of the triangle representing a different domain: the macroscopic domain (properties observable by the human senses), the sub-microscopic or particulate domain (the structures of atoms and molecules that give rise to their properties), and the symbolic domain (the symbols, equations, and abbreviations that compose the "foreign language," if you will, of communicating chemistry). Johnstone argued one reason that students found chemistry so difficult to learn was that it was taught and tested almost exclusively in the symbolic domain with few efforts to make concrete, explicit connections to the other two domains.

Even though teachers did demonstrations and students carried out lab experiments, assessment focused on mastering the symbols, equations, and mathematics of chemistry. A collection of papers published in the early 1990s in the *Journal of Chemical Education* demonstrated that even when students were successful with assessments in the symbolic domain, they struggled in the particulate domain (*16–20*). Students could solve stoichiometry problems through a variety of mathematical manipulations (factor-label, dimensional analysis) because they had memorized the procedure, or algorithm, for doing so. When asked to solve an analogous problem depicting reactants and products as particles rather than giving the mass of reactants and asking for the mass of products, students were much less successful. The papers by Nurrenbern, Pickering, and others presented data from multiple studies that students struggled to think about the concepts underlying stoichiometry. Chemistry teachers could no longer ignore the reality that many students performed well on assessments, yet lacked understanding of key concepts. Rather, students often performed well because they had memorized facts and algorithms. Teachers could no longer presume that correct answers on assessments were sufficient evidence that students understood the ideas, concepts, principles, and theories of chemistry.

A Watershed Year for Assessment and Chemistry Education: 1994

In 1994, chemistry education research was showcased in four high profile events, each of which contributed to the heightened importance of assessment in chemistry education. First, the National Science Foundation funded 14 planning grants for the Systemic Initiatives in Chemistry in 1994 (*21*). Assessment became an essential component, and defining characteristic, of the reforms that were fully funded (the ChemLinks and ModularChem Consortium, Molecular Science, the New Traditions Project, and Workshop Chemistry). In addition to challenging the paradigms of what chemistry should be taught and how, the Systemic Initiatives also provided unprecedented professional development opportunities to build capacity in assessment. Multiple chemistry education research (CER) scholars were post-doctoral research scholars as part of the systemic initiatives, including

Dr. Jennifer Lewis (University of South Florida), Dr. Dawn Rickey (Colorado State University), and Dr. Stacey Lowery Bretz (Miami University).

Faculty involved in the planning grants participated in a second exceptional event just days later. Art Ellis and Angelica Stacy co-chaired the first Gordon Research Conference focused exclusively on chemistry education, Innovations in College Chemistry Teaching (*22*). [The conference changed its name to "Chemistry Education Research and Practice" in 2003.] The program for this conference included multiple talks focused explicitly on testing for measurement vs. understanding and showcased several curricular and pedagogical experiments.

A third significant event that year was the publication of the Task Force on Chemistry Education Research report (*23*) that characterized the similarities and differences between bench chemistry research and CER. The report highlighted that CER share many similar goals, such as elucidating mechanisms (of teaching and learning), identifying intermediates (along the path to learning), synthesizing new materials (to increase learning), and characterizing products (of learning). Even more importantly, the report cautioned chemists about some key differences with regard to measurement and assessment. For example, research designs that hold all independent variables constant while manipulating just one are both ethically and logistically problematic when conducting human subjects research.

The Task Force report paved the way for the fourth significant event that year – a symposium on the methods used to conduct CER (*24*). Papers presented important methodological considerations for both quantitative (*25*) and qualitative research (*26*). Techniques for assessment of student learning were featured, including think-aloud interviews (*27*) and surveys and questionnaires (*28*).

Assessment Design as Research: 2000 and Beyond

As assessment became an important "feedback loop" for experiments in the systemic initiatives, chemists began to focus on the design and validation of new tools specific to chemistry. Mulford and Robinson's Chemical Concept Inventory (*29*) featured multiple items regarding student misconceptions about the particulate nature of matter. Assessments also began to broaden beyond just measuring content knowledge, or the lack thereof. The Chemistry Laboratory Anxiety Instrument (*30*) was the first assessment tool specifically designed to look at the dimensions unique to learning within the chemistry laboratory environment. The Chemistry Attitudes and Experiences Questionnaire (*31*) was developed to measure first-year university chemistry students' attitudes toward chemistry and their self-efficacy in order to investigate the factors that influence students' enrollment choices. All three of these publications advanced the methods of assessment development in chemistry by including discussion of protocols for establishing the reliability and validity of the data generated by the instruments.

CER scholars once again turned to learning theory to make data-driven choices about what to assess as evidence of learning and why. Perry's model of intellectual and ethic development amongst college students was introduced to chemists (*32, 33*). The *Journal of Chemical Education* published its first "online symposium" in 2001 featuring theoretical frameworks including George

Kelly's constructivism (*34*), Andrea diSessa's naïve ideas (*35*), Joseph Novak's meaningful learning (*36*), David Kolb's experiential learning (*37*), David Mezirow's transformative learning (*38*), knowledge frameworks (*39*), and Piaget's theory (*40*, *41*). Each manuscript discussed what ought to be assessed given its theoretical stance on how learning takes place.

Discipline-specific assessment tools were so critical to education reforms that in 2001, the National Science Foundation created a new track within the Course, Curriculum, and Laboratory Improvement program called "Assessment of Student Achievement" (ASA). The request for proposals described the purpose of the ASA track as "support[ing] research on assessment and the development and dissemination of assessment practices, materials, and measures to guide efforts to improve the effectiveness of courses, curricula, and programs of study." Four ASA awards were made within chemistry: CHEMQuery (*42*), CHEMX (*43*), IMMEX (*44*), and LUCID (*45*). In 2007, proceedings from a conference featuring all the ASA awards were co-published by the National Science Foundation and Drury University (*46*).

Despite content knowledge being traditionally taught and tested one course at a time, the American Chemical Society Examinations Institute released a unique assessment tool (*47*) specifically designed to examine the ability of undergraduate students to integrate knowledge across all their chemistry courses. The Diagnostic of Undergraduate Chemistry Knowledge (DUCK), which requires data analysis in the context of interdisciplinary scenarios, is intended to be administered at the end of an undergraduate degree.

As additional CER scholars developed a particular expertise in assessment, the conversation shifted from gathering data about the outcome of a course (or curriculum) to gathering data about what's happening in the course (*48*). Multiple tools have since been published to generate such data regarding students' metacognition and affective learning, including the Chemistry Self-Concept Inventory (*49*), the Metacognitive Activities Inventory (*50*), and the Attitudes toward the Subject of Chemistry Inventory (*51–53*).

Additional assessment tools have recently been developed that focus on measuring students' understandings of structure and multiple representations. For example (*54*), developed the Implicit Information from Lewis Structures Instrument to assess what connections students draw between structure and properties. Linenberger and Bretz (*55*) developed an interview protocol to reveal students' cognitive dissonance about multiple representations and created the Enzyme-Substrate Interactions Concept Inventory to measure their understanding of such multiple representations (*56*).

Future Frontiers in Assessment

While the scholarship of assessment in chemistry education has grown significantly in the last century, challenges remain. For example, the "concept learning vs. problem solving" research of the early 1990s focused heavily on assessing students' understanding of particulate images (*16–20*). While some might consider that literature to have definitively answered the debate about "conceptual vs. algorithmic" understandings, the practice of teaching and learning chemistry have changed significantly since that time. Twenty years later, particulate images are published in multiple colors in textbooks and access to computers has completely changed the landscape of teaching and learning with visual representations in chemistry. The ACS Exams Institute released two general chemistry exams (2005 & 2007) that purposefully paired conceptual and traditional items. Contrary to the findings in the early 1990s, analysis of data from over 3500 students on these paired-item exams indicates that there was *no difference* in performance when comparing the conceptual and traditional items (*57*). Chemistry teachers are no longer limited to using only static images to assess particulate understanding given the plethora of animations and simulations. There is limited evidence that animations of particle motion actually produces different results (*58*) than those in the early 1990s.

Clearly, once an assessment has been developed, validated, and published does not mean it will be valid for all students and all classrooms going forward. Each teacher or researcher bears the responsibility to judge the validity of an assessment in his or her particular context, including attending to changes in the content and context of teaching and learning chemistry since the tool was originally developed. A recent review by Arjoon et al examines current practices in chemistry education with regard to psychometrics and the design of assessments (*59*).

Another challenge facing chemists with regard to assessing student learning is how much is too much? While the development of assessment tools to measure multiple dimensions of cognitive and affective learning in chemistry offers many options to the chemist studying curricular or pedagogical changes, the very real issue of "assessment fatigue" (*60*) looms from the perspective of the student. The indiscriminate measuring of any and all aspects of student learning yields no useful formative or summative data to improve teaching and learning, but runs the danger of students who respond haphazardly as they tire from answering one assessment after another. Rather, chemistry teachers and chemistry education researchers would be well advised to make judicious choices about what knowledge or attitudes or skills are of particular interest in a specific context in order to focus upon measuring those. The alignment between the measurement and the knowledge/attitudes/skills is just as essential in the assessment of student learning as it is in the chemist's laboratory when assuring alignment between the signal and the detector in any instrument.

References

1. Cornog, J.; Colbert, J. C. *J. Chem. Educ.* **1924**, *1*, 5–12.
2. Powers, S. R. *J. Chem. Educ.* **1924**, *1*, 139–144.
3. ACS Exams, 2013. http://chemexams.chem.iastate.edu/about/short_history.cfm (accessed February 18, 2013).
4. Strong, L. E.; Wilson, M. K. *J. Chem. Educ.* **1958**, *35* (2), 56–58.
5. Merrill, R. J.; Ridgway, D. W. *The CHEM Study Story*; W.H. Freeman: San Francisco, 1969.

6. Davenport, D. A. *J. Chem. Educ.* **1970**, *47* (4), 271.
7. Herron, J. D. *J. Chem. Educ.* **1975**, *52* (3), 145–150.
8. Herron, J. D. *J. Chem. Educ.* **1978**, *55* (3), 190–191.
9. Driver, R. *Eur. J. Sci. Educ.* **1981**, *3*, 93–101.
10. Linke, R. D.; Venz, M. I. *Res. Sci. Ed.* **1978**, *8*, 183–193.
11. Linke, R. D.; Venz, M. I. *Res. Sci. Ed.* **1979**, *14*, 78–88.
12. Osborne, R. J.; Cosgrove, M. M. *J. Res. Sci. Teach.* **1983**, *20*, 824–838.
13. Peterson, R. F.; Treagust, D. F.; Garnett, P. *Res. Sci. Ed.* **1986**, *16*, 40–48.
14. Treagust, D. F. *Int. J. Sci. Educ.* **1988**, *10*, 159–169.
15. Johnstone, A. H. *J. Comput. Assist. Learn.* **1991**, *7* (2), 75–83.
16. Nurrenbern, S. C.; Pickering, M. *J. Chem. Educ.* **1987**, *64*, 508–510.
17. Pickering, M. *J. Chem. Educ.* **1990**, *67*, 254–255.
18. Sawrey, B. A. *J. Chem. Educ.* **1990**, *67*, 253–254.
19. Nakhleh, M. B. *J. Chem. Educ.* **1993**, *70*, 52–55.
20. Nakhleh, M. B.; Mitchell, R. *J. Chem. Educ.* **1993**, *70*, 190–192.
21. Hixson, S. H.; Sears, C. T., Jr. *J. Chem. Educ.* **1994**, *71*, 43–46.
22. Gordon Research Conferences, 2013. Innovations in College Chemistry Teaching. https://www.grc.org/programs.aspx?year=1994&program=0000901 (accessed February 18, 2013).
23. Bunce, D. M.; Gabel, D.; Herron, J. D.; Jones, L. *J. Chem. Educ.* **1994**, *71*, 850–852.
24. Metz, P. A. *J. Chem. Educ.* **1994**, *71* (3), 180–181.
25. Nurrenbern, S. C.; Robinson, W. R. *J. Chem. Educ.* **1994**, *71* (3), 181–183.
26. Phelps, A. J. *J. Chem. Educ.* **1994**, *71* (3), 191–194.
27. Bowen, C. W. *J. Chem. Educ.* **1994**, *71* (3), 184–190.
28. Pribyl, J. R. *J. Chem. Educ.* **1994**, *71* (3), 195–196.
29. Mulford, D. R.; Robinson, W. R. *J. Chem. Educ.* **2002**, *79* (6), 739–744.
30. Bowen, C. W. *Educ. Psychol. Meas.* **1999**, *59*, 171–185.
31. Dalgety, J.; Coll, R. K.; Jones, A. *J. Res. Sci. Teach.* **2003**, *40* (7), 649–668.
32. Finster, D. C. *J. Chem. Educ.* **1989**, *66* (8), 659–661.
33. Finster, D. C. *J. Chem. Educ.* **1991**, *68* (9), 752–756.
34. Bodner, G. M.; Klobuchar, M.; Geelan, D. *J. Chem. Educ.* **2001**, *78* (8), 1107.4.
35. Nakhleh, M. B. *J. Chem. Educ.* **2001**, *78* (8), 1107.5.
36. Bretz, S. L. *J. Chem. Educ.* **2001**, *78* (8), 1107.6.
37. Towns, M. H. *J. Chem. Educ.* **2001**, *78* (8), 1107.7.
38. Wink, D. J. *J. Chem. Educ.* **2001**, *78* (8), 1107.8.
39. Robinson, W. R.; Samarapungavan, A. *J. Chem. Educ.* **2001**, *78* (8), 1107.3.
40. Bunce, D. M. *J. Chem. Educ.* **2001**, *78* (8), 1107.2.
41. Nurrenbern, S. C. *J. Chem. Educ.* **2001**, *78* (8), 1107.1.
42. Scalise, K. Ph.D. Thesis, University of California, Berkeley, Berkeley, CA, 2004.
43. Grove, N. P.; Bretz, S. L. *J. Chem. Educ.* **2007**, *84* (9), 1524–1529.
44. Cox, C. T., Jr.; Jordan, J.; Cooper, M. M.; Stevens, R. *Sci. Teach.* **2006**, *73*, 56–60.
45. Wolfskill, T.; Tofan, D.; Fong, A.; Hanson, D. *LUCID: A Web-Based Learning and Assessment System*; CONFCHEM Online Conference

on Non-Traditional Teaching Methods, April 21–22, 2003. http://www.files.chem.vt.edu/confchem/2003/a/lucid/lucidtwdh.htm (last accessed February 18, 2013).

46. Deeds, D.; Callen, B., Eds.; *Proceedings of the National STEM Assessment Conference*; National Science Foundation and Drury University, 2008.
47. Holme, T. A.; Murphy, K. L. *Diagnostic of Undergraduate Chemistry Knowledge Exam*; Division of Chemical Education Examinations Institute, American Chemical Society: Washington, DC, 2008.
48. Bauer, C. F.; Cole, R. S.; Walter, M. F. In *Nuts and Bolts of Chemical Education Research*; Bunce, D. M.; Cole, R. S., Eds.; ACS Symposium Series 976; American Chemical Society: Washington, DC, 2008; pp 183–201.
49. Bauer, C. F. *J. Chem. Educ.* **2005**, *82* (12), 1864–1870.
50. Cooper, M. M.; Sandi-Urena, S. *J. Chem. Educ.* **2009**, *85*, 240–245.
51. Bauer, C. F. *J. Chem. Educ.* **2008**, *85* (10), 1440–1445.
52. Brandriet, A. R.; Xu, X.; Bretz, S. L.; Lewis, J. *Chem. Educ. Res. Pract.* **2011**, *12*, 271–278.
53. Xu, X.; Lewis, J. E. *J. Chem. Educ.* **2011**, *88*, 561–568.
54. Cooper, M. M.; Underwood, S. M.; Hilley, C. Z. *Chem. Educ. Res. Pract.* **2012**, *13*, 195–200.
55. Linenberger, K. J.; Bretz, S. L. *Chem. Educ. Res. Pract.* **2012**, *13*, 161–171.
56. Bretz, S. L.; Linenberger, K. J. *Biochem. Molec. Biol. Educ.* **2012**, *40* (4), 229–233.
57. Holme, T. A.; Murphy, K. In *Investigating Classroom Myths through Research on Teaching and Learning*; Bunce, D. M., Ed.; ACS Symposium Series 1074; American Chemical Society: Washington, DC, 2011; pp 195–206.
58. Sanger, M. J.; Campbell, E.; Felker, J.; Spencer, C. *J. Chem. Educ.* **2007**, *84* (5), 875–879.
59. Arjoon, J. A.; Xu, X.; Lewis, J. E. *J. Chem. Educ.* **2013**, *90* (5), 536–545.
60. Holme, T. A.; Bretz, S. L.; Cooper, M. M.; Lewis, J.; Pienta, N.; Stacy, A.; Stevens, R.; Towns, M. H. *Chem. Educ. Res. Pract.* **2010**, *11*, 92–97.

Trajectories of Reform of Institutions and Institutional Collaboration

Chapter 11

Lessons Learned from Collaborations in Chemistry Assessment across Universities: Challenges in Transfer and Scale

Pamela L. Paek[1] and Thomas A. Holme[*,2]

[1]Center for Assessment, Austin, Texas 78733
[2]Department of Chemistry, Iowa State University Ames, Iowa 50011
[*]E-mail: taholme@iastate.edu

This chapter reviews a collaborative effort to cross-pollinate and share work around chemistry assessments across several universities. The goal was to find ways to synthesize separate projects and capitalize on applying developed instruments and assessments beyond a single university, and in new situations, to increase scale and check for generalizability. Discussion of the successes and challenges of scale and transfer of the collaboration is detailed in this chapter.

By definition, collaboration is "a coordinated, synchronous activity that is the result of a continued attempt to construct and maintain a shared conception of a problem (*1*)." In theory, collaborations appear to be an easy way to combine the power of multiple minds in the joint effort of developing a product or set of products that could not be done by a single individual (*1*, *2*). However, in practice, there are logistical issues (*3*), different mindsets and habits of mind (*4*), as well as unspoken end goals or motivations (*5*) that can impact the effect collaborations can have. Additionally, underestimating these issues moderates the amount of transfer and scale that is possible in multi-site collaborations (*6*, *7*).

This chapter discusses the synthesis of the multiple projects across multiple universities to address a larger issue in undergraduate chemistry. In this collaboration, the goal was to develop a system of chemistry assessments that could be used collectively to inform instruction for undergraduate chemistry instructors. The thought was that the combined use and comparison of different

© 2013 American Chemical Society

measures would provide insight into how the instruments better determined similar or different types of knowledge and understanding than a single project usually would entail. This collaborative effort synthesized work together by having separate projects interact. It was not just a summary of different discrete projects, rather, a variety of combinations that would allow projects to interact in various ways, within and across different universities, chemistry courses, and combinations of uses.

The outcomes of the overall collaborative effort, both in terms of the successes and challenges, will be examined through five main themes: (1) previous partnerships prior to this large-group collaboration, (2) similarities and differences in project goals, (3) university support for conducting and scaling up chemistry education research, (4) the use of original and modified instrument measures for comparison across years, classrooms, and universities and (5) how instructor beliefs and values affect how they use different information for assessing student learning. The first two ideas are grouped under the theme of collaboration, while the latter two themes focus on scale and transfer. The third idea about university support, deals with both collaborations as well as scale and transfer.

Collaborations: Previous Partnerships Prior to This Large-Group Effort

This collaborative effort consisted of eight principal investigators from eight different universities. Each of these PIs is a distinguished faculty member specializing in chemistry education, all with a focus on synthesizing assessment ideas and instruments within undergraduate chemistry. They are listed below in alphabetical order with their department and current university affiliation:

- Stacey Lowery Bretz, Department of Chemistry & Biochemistry, Miami University
- Melanie Cooper, Department of Chemistry, Michigan State University
- Thomas Holme, Department of Chemistry, Iowa State University
- Jennifer Lewis, Department of Chemistry, University of South Florida
- Norbert Pienta, Department of Chemistry, University of Georgia
- Angelica Stacy, Department of Chemistry, University of California, Berkeley
- Ronald Stevens. Department of Microbiology, Immunology, and Molecular Genetics, University of California, Los Angeles
- Marcy Towns, Department of Chemistry, Purdue University

As proof of concept to show collaborations yielded successful results, all eight principal investigators (PIs) on the larger project worked together in some capacity – mainly through pairwise partnerships – previous to the award of this collaborative grant through the National Science Foundation's Course, Curriculum, and Laboratory Improvement (CCLI) grants: Collaborative Research: A Model for Data-Driven Reform in Chemistry Education (Award ID

DUE-0817409). One of the reasons NSF was interested in supporting this larger collaborative effort was because of the previous successes of the PIs working together in pairs or triads, showing some initial synergies for advancing their work in chemistry education. Another was that each of the principal investigators was in their own right leaders in chemistry assessment, and as a group they might contribute to a new wave of how these assessments and corresponding information could be used to improve instructional practices of undergraduate chemistry instructors.

What is interesting to note is that none of the PIs were from schools of education, rather, they were housed in departments of chemistry, or in one case, a different, but closely related, science department. So, the fact that they were all based in their strong content knowledge of chemistry, yet interested in employing educational practices like development and measurement of content-based and non-cognitive assessments to their content area, was indeed a major undertaking and leap forward to advance how assessments could inform undergraduate chemistry instruction. In short, chemistry faculty members specializing in chemistry assessment is new area of study that currently only has a small number of researchers invested in the work. This group wanted to help support chemistry instructors by better understanding what undergraduate chemistry students understood and had misconceptions about from a content perspective. From a non-content cognitive perspective, the group was also interested in developing and using measures of students' self-efficacy and metacognitive skills, which they postulated could be used to target students more at risk because of a lack of confidence or lacking of study skills to effectively succeed, primarily in a first-semester, gateway, chemistry course.

Collaborations: Similarities and Differences in Project Goals

Given the similar outcomes and intent of previous individual and pair-wise partnerships, the collaborative effort of eight principal investigators along with similar research interests for measuring student cognitive and non-cognitive factors, the project envisioned collaborations to expand both the depth of the assessment research and the scale of the application of assessment instruments. Thus, the overarching interest of the collaborators was ultimately to both improve chemistry instruction and develop students' metacognitive skills. From the outset, all partners in this project seemed to have the same overall goal, where each PI would be able to contribute his/her part. An outside evaluator with specialization in developing and evaluating educational assessments, Pamela Paek, was charged with analyzing the findings of this collaborative effort. This chapter demonstrates this evaluation, along with recommendations for future collaborations.

Collaboration Proposal and Plans

The collaboration was initially proposed as a three-year study, to provide time to conduct synergistic activities beyond single universities across multiple years in chemistry assessment. This format would allow PIs to replicate findings over multiple years, potentially begin some longitudinal studies, and allow for adaptations and modifications as needed to extend or refine current instruments for further study and analysis. However, the funding level for the project was notably less than what was initially proposed and the timeframe was reduced to 18-months. As such, the project's change of scope focused more on the initial activities that could be done across universities, but not necessarily gather data beyond a single year or be able to move too far with scale and transfer, given the timeframe.

As in any research project, an 18-month research study involving students and faculty would mean potentially only 2 semesters worth of data, assuming each researcher would be able to have research instruments ready to administer, human subjects approval, and other logistics in place for other instructors on board to implement, as well as graduate students to support the work. While this 18-month funding window in principle allowed a basis for initiating collaborations across universities, the logistics for doing so required substantially more time to successfully carry off within each institution, and was compounded when trying to scale beyond a single university and researcher.

Since all collaborators had been previously involved in using and developing different assessment instruments, the goal of the first six months was to use existing instruments to gather baseline data, provide comparative data for students and instruments at the institutions involved in the project, and address logistical issues that may arise from the collaboration. The challenge here was not about continuing research on work that was already in place, but around the collaborative efforts to cross-pollinate assessment instruments and conduct comparative studies between institutions, and across cohorts within institutions. This challenge proved to be the most difficult, outside of the issues of faculty members within institutions not utilizing the instruments for the purpose of informing instruction. So, scale became problematic not only within universities beyond the main researcher onsite, but also across universities.

The researchers were realistic in that the limited timeframe would inhibit scope, as they would be able to establish baseline data, but not be able to conduct longitudinal studies. There would also be limited opportunities for repeating studies to confirm initial observations related to content changes for assessments that were still being developed and refined. Additionally, the ability to discover several synergies between assessment instruments was limited due to how quickly each researcher was able to pool resources to utilize collaborators' instruments and integrate those with their own instruments at each university site.

While the group requested and was approved for a no-cost extension of this grant, the time that was afforded was merely to provide setup for the activities they wanted to engage in. The original timeframe was extended to allow for more time on instrument development, plans for cross-site studies, and initial analyses of these results. Even with the more limited funds acquired via this grant, with the extended time, the PIs were able to produce a myriad of presentations and publications that highlighted the different partnerships in this collaboration, one of which was a joint publication of all eight PIs (*8*).

Refocused Collaborative Efforts in a Compressed Timeframe

A large focus of the collaboration was on development and uses of affective and metacognition measures, as seen in the use of the full Metacognitive Awareness Inventory (MCAI) (*9*, *10*), a modified version of the MCAI, a modified version of the Attitude toward the Subject of Chemistry Inventory (ASCI) (*11*, *12*), and CHEMX (*13*, *14*). The goals of using these measures were to see how performance on these assessments related to students' performance on chemistry assessments. Data collection included gathering pre-test performance to compare to later chemistry assessment performance as well as gathering post-test performance to evaluate change on the non-cognitive measures and how this change was related to understanding of chemistry content. The goal was to analyze the relationship of these data points and to see how instructors may be able to use the pre-test measures as ways to intervene and better support students' efforts and approaches to learning.

A second area of focus was on developing ways to better assess students' approaches to learning, as evidenced by the use of an instrument measuring key concepts (*15–17*) that include the misconceptions students may have on these complex topics (*18*, *19*), reasoning concepts using the Test of Logical Thinking (TOLT) and Group Assessment of Logical Thinking (GALT) (*20*), as well as the use of IMMEX (*21*, *22*). The goal of using these measures was to find a way to better understand students' problem solving strategies and key areas of misconceptions, to help instructors have the data they need to support students and target instruction more effectively. Figure 1 provides a timeline of the various implementation of instruments over the various schools in the project. Schools appearing above the arrows indicate the location where the instrument was developed or adapted, and those below the arrows are the additional schools the used that instrument and thereby provided cross-validation data available to this project.

While there were similarities and interests in the different assessments developed or under development, there was no whole group effort to use one set of assessments across all eight universities. Rather, the collaboration appeared to be a continuation, with some expansions, of previous partnerships that extended into new and refined instruments, or extending the pairs into groups of three or four. The one common thread for several of these small groupings was the collaborator who focused on the measurement effort, as her role was to analyze each new assessment to validate the overall construct. This included, for example, investigating factor loadings, and providing advice on possible refinements, using statistics, for future versions of each assessment. This activity, however, did not extend to each collaborative implementation of assessment carried out within the project.

Even with the compressed timeframe and complexities of conducting a multi-PI research endeavor, a significant amount of collaborative research was conducted, with results published of these collaborative efforts in twelve publications, including one joint publication of all principal investigators and the evaluator of this collaborative (*8*). The whole group publication described the intent of the collaboration and the goals for working across universities, instruments, and contexts, to show how these different measures could work synergistically to move the chemistry assessment field forward and assist in the way chemistry instructors could improve their teaching, and thus increase the success of their students' master of core content in their classes. Funding from the grant was often used to support graduate students within the individual PI research groups, so seven of the publications from the project include graduate student co-authors.

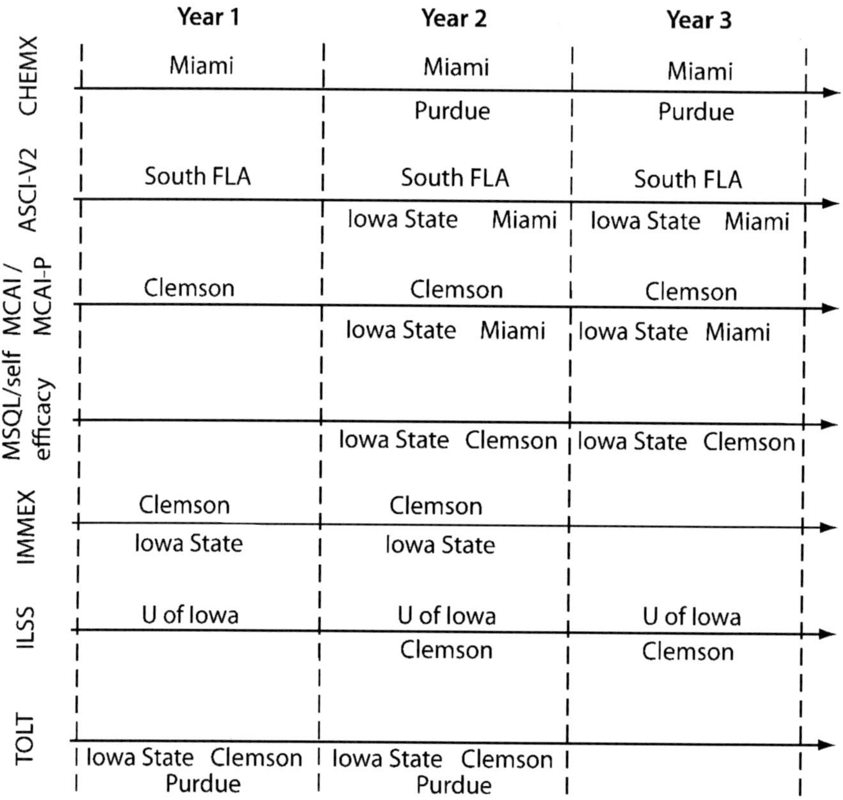

Figure 1. Timelines for implementation of assessment instruments at collaborating universities. Instruments are listed down the left side. A university listed above the arrow is the instrument developer and those listed below the arrow are instrument users.

It is an understatement to say collaborations take considerable energy and effort to keep everyone focused on the same sets of goals and outcomes. Two meetings of all PIs on the project were held in order to enhance communication and re-establish group priorities during the project. Nonetheless, with more time and funding to support the diversity of expertise and opinion, this collaboration could have been even more successful in terms of the increased transfer and scale across sites of the multiple instruments that have been developed. While quite a bit of progress was made, interviews with each of the PIs indicated their desire for more years of data to study. As such, both increased time and funding were needed to harness the overall strength of this group of researchers, to make the impact they wanted to make collectively, which they only began to do with this initial grant.

University Support for Conducting Research

Each collaborator was very thoughtful and inclusive in the way they included their graduate students in their research. The number of publications that included graduate students as coauthors testifies to the fact that these PIs knew how to mentor and support their students. In many cases, the students were given autonomy to lead parts of their research projects, as well as publish not only as secondary authors, but first authors of the research. If left to do their own research without a need to transfer and scale their work to other instructors within their department, each PI demonstrated high success in their independent endeavors, as well as through their students. It was only when the collaborators had to rely on faculty peers or college or university administrative efforts that their research agenda was compromised: by lack of buy-in, support, and investment from other faculty members, and administration to truly use the developed measures as they were intended.

When collaborators would ask other chemistry instructors to administer their assessments and use the data to inform their practice, the results from these peers were always lower than the partners in the collaborations. These findings may be a result of faculty peers not understanding the benefits as clearly as the PIs for using the information. These cooperating faculty members may also not have shared the same level of interest for improving their instruction. So from the outset, different motivations for using the assessments lead to different implementation and use of the information. In fact, two of the PIs specifically analyzed instructor's perceptions of what content they prioritized, and the types of information they used to support those beliefs (*23*). This is discussed further in the section below related to transfer and scale.

The other challenges that assessment studies often face arise from a need for student level data beyond what a PI may automatically have for students in their course. For instance, data such as previous science performance, GPA, demographic information, or other data that could be used to adjust for prior performance, or be used to demonstrate potential differences by subgroups is often unavailable or quite difficult to obtain. For courses taught by faculty members outside of the collaboration, PIs were either unable, or significantly delayed (more than two years at one university) to gain access to such data, which was critical for their analyses to demonstrate similarities and differences of classes, and to even make headway for generalizing results. Without this information, it was impossible to begin to study how well assessments scaled or transferred when used by other faculty members. This type of hold-up obviously affects the timeliness of research for individuals or groups to be published, but also demonstrates the lack of support mechanisms at universities to provide discipline-based education research (DBER) faculty members the data they need to conduct their research. Given that all universities participating in the current project have top tier research ratings, one would think that administrative efforts to facilitate research would be more universal for DBER faculty members, allowing them to contribute to the university's reputation as strong research institutions. This comment is not to say that the collaborators were directly blocked from access to additional data. Rather, this observation reflects an overall sense of institutional apathy towards being more proactive to support their faculty researchers. It is the lack of action and/or attention of university administration that is the issue here. With less tenacious individuals, the amount of publications and research that would have resulted would likely have been slim or even none. The role of institutional barriers to successful scale-up of assessment research represents a key finding of this analysis of large scale, cross-institution collaborations, even though the finding is a result within what may broadly be considered research university environments. The point to be made is that until individual university administrations are more proactive and supportive of the data they can provide faculty research, collaborations across universities will be further hindered in their ability to transfer and scale research.

Transfer and Scale: The Use of Original and Modified Instrument Measures for Comparison Across Years, Classrooms, and Universities

McDonald and colleagues define scale as "the practice of introducing proven interventions into new settings with the goal of producing similarly positive effects in larger, more diverse populations (*7*)." Part of scale includes modification and transfer, where initial research provides information to potentially improve and refine initial measures and hypotheses, and then replicate the results within similar settings or scale to other settings.

One premise of the members of the collaboration was that chemistry instructors are more likely to adopt assessment instruments that require little time to use and analyze. As such, early on in the project, a shortened version of a previously published instrument (*24*) was proposed and validated (*11*). This new ASCI instrument as well as a modified version of MCA-I were then tested across three universities, to see what findings would hold or differ across sites and demographics.

Similarly, a preliminary study of CHEMX across these three universities was conducted, to analyze how well results generalized within and across sites (*14*). While there were the hopes for the overall project to carry out work of this nature across more instruments, including more years within and across universities, in the timeframe available, only these smaller studies were achievable. However, these studies showed promise for how this information could be used to inform instructors about cognitive factors, both content and non-content related. These findings would then need to be incorporated by instructors in how they would use such data to change their practice, which is a point of further study for this team of researchers. This collaboration was only on the verge of exploring generalizability and practical issues around implementation to study transfer and scale. This broader implementation step is where the next issue of transfer and scale comes into play.

Transfer and Scale: How Faculty Beliefs and Values Impact How Different Information Is Used for Assessing Student Learning

Within any academic discipline there are two main approaches for dealing with content coverage within a course: treating a broad range of topics rather lightly or addressing fewer topics in greater depth. A significant amount of material can be covered at a relatively superficial level while more integrated in-depth exploration of fewer topics may mean deeper understanding of a smaller range of content. Choices made in this regard have a large impact on assessment choices that accompany instruction. This project was implemented largely within introductory chemistry courses. Because these courses include a large and growing list of topics, with high expectations of mastery, coverage that balances breadth and depth is a constant challenge for instructors of these courses. As Cooper states, "general chemistry... covers too much material, thereby sacrificing depth for breadth (*25*)."

What compounds the problem of depth versus breadth is instructors' understanding of how students learn. A relative lack of familiarity often leads to the failure to use instructional strategies that would engender more student motivation and interest as well as sound pedagogical techniques for ensuring mastery of content. While not universally true, chemistry faculty members at research universities may treat the teaching of undergraduate courses as a less attractive part of their academic responsibilities. Therefore, it is unwise to assume that instructors in the large-lecture introductory courses are particularly interested in the ways students learn. It would also be a notable assumption that many of them have a profound understanding of educational assessment. Operationally, it appears more than likely that they are interested in instruments that survey a wide host of concepts (hence, wider breadth) rather than depth. Such assessment is in line with faculty beliefs that the purpose of these introductory courses is generally to provide an overview of the subject. In fact, an article focusing on the questions to ask instructors about assessment, not just as a compliance task, but actually making meaning of what assessments can do (*26*), demonstrates one form of professional development that could help research faculty make more meaning out of assessment efforts in introductory courses. Ideas such as this help frame the reasons for why assessment development work is important. In addition, research that investigates the reasons instructors enumerate as to why change may not be happening becomes vital because of different beliefs and values of what teaching is, and what is important to learn (*23*).

The role of assessment within higher education is not wholly ignored, but the willingness of many instructors to commit limited time resources to enhancing their measures of student learning is apparent. This situation may have a particularly large effect in science courses that occupy a service role in the curriculum of a majority of the enrolled students. Additionally, faculty members who lack strong foundations or interest in instructional methods may inadvertently eliminate potential chemistry majors because they did not engage all students to be successful in learning the content of these general chemistry courses. The research that the collaborators in this project conducted on misconceptions, reasoning, conceptual frameworks, and problem solving all demonstrated that if this information was not somehow attended to, students would generally lack a true understanding of complex ideas. Without such a depth of knowledge these students can be expected to be less successful at carrying the content information into future chemistry courses (if they continued). They would also miss out on opportunities to improve the way they could reason and problem solve, because they were not given more opportunities to improve upon those higher-order thinking skills. In short, without using assessment information of the type provided by this collaboration, other faculty continue to miss opportunities to help students learn the content more deeply, improve upon their general approaches to learning—including self-efficacy and metacognition—and overall, inadvertently contribute to the attrition of science majors. Thus, the challenges with obtaining buy-in from fellow faculty members that the collaborators faced in this project become a particularly important observation. *Even with high quality, publishable results from DBER efforts, the transfer of these ideas to other instructors represents a central challenge in the cause of using sound evidence of student learning in the reform of teaching and learning.* Not only does this impact the teaching efforts of instructors, but in the longer term serves to limit the pool of new science talent because students are less engaged and instructors are less focused on improving the ways their students can learn and grow. It will be important that future grants include time and money to support faculty buy-in and professional development to ensure more success in transfer and scale of reform efforts.

Summary

Ultimately, the yet unachieved goal of this collaborative was to point to a new era within chemistry assessment. Not only can there be measures to improve the ways different assessments can unpack students' misconceptions, knowledge, and interest, they can also inform instruction, to assist more students to achieve success in chemistry. And further, better understanding the expectations of faculty members informs the fledgling chemistry assessment research in what these instructors see as critical to assess and teach. These studies of the academic environment provide more insight into why certain measures or data would not be used or disregarded to inform their practice. As such, headway is being made, but more time and research is needed to forge ahead with a new era of new chemistry assessment use.

For change to happen within and across each educational level, the structures and support processes must be revised to accommodate and successfully implement (as well as sustain) large-scale change across all levels. This collaboration suggests that for scale-up, there is a need for both buy-in and support from instructors as well as administration to successfully implement these reform efforts. Additionally, it is important to understand the contexts in which faculty members operate and to better understand their own beliefs about student learning and the types of data they value for measuring student learning. Without a better understanding of sources of resistence to the use of assessment instruments, scale-up will prove challenging, especially if results vary due to faculty beliefs and values of how and what students should learn in these undergraduate chemistry courses. This point must not be taken lightly – as this issue is systemic in education, from K-12 to higher education.

For the assessment and evaluation community, there are certainly significant issues related to scaling and transferring research to more practical venues. Because of the successful and perseverant nature of each of these collaborators on this project, they were able to champion their own work, move forward with their own research agendas, and still individually be quite prolific in their publications. The progress of STEM education reform cannot become solely dependent on the resilience and persistence of individuals for research to be successful. Rather, the development of infrastructure that would enable participants to be successful in collaborative efforts becomes apparent, and a must-needed investment. Several different levels of institutional involvement are important, in particular: (1) university support from both peers and administration; (2) grant funding that provides adequate resources and opportunities for collaborators in various locations to work together onsite and over longer time scales; (3) resources to hire a person in charge of coordinating separate endeavors as their sole responsibility; and (4) an understanding from funding agencies about how long it takes to develop, cultivate, and support collaborations. There is little that can be done in a short timeframe such as 18 months, and for scale and transfer to happen, more time, money, and logistical support is needed to provide any opportunities to replicate results over multiple years across sites. Collaborators on projects of this type need to spend more time to cross-pollinate and learn from each others' efforts and results, and continue to build and refine how each participant's efforts contributes to the overall project. As shown in many individual projects, this work can be done if there is one PI and one project as the focus. There has been relatively little work, however, that attempts to understand collaborative efforts that arise from multiple individual PI projects. This is definitely a missed opportunity to synthesize individual work into something larger that has such potential for positive change to undergraduate science teaching and learning.

This chapter identified some characteristics of collaborative projects that funding agencies can use to discern probable success and to learn what it would take to help those project do well. To summarize, the need for grants that span a longer period, similar to those in scale-up studies would be most appropriate for any collaborative effort, as collaboration can be viewed as a version of scale —with slight modifications or development, rather than full scale-up and efficacy trials. If collaborations—those where PIs combine their research agendas into a larger set of studies, not just working jointly on a project within or across departments and/or universities—could be classified in their own category, that would also prove helpful to address the complexity of what it takes to truly cross-pollinate multiple research projects.

References

1. Hafernik, J. J.; Messerschmitt, D. S.; Vandrick, S. Collaborative research: Why and how? *Educ. Res.* **1997**, *26*, 31–35.
2. Wasser, J. D.; Bresler, L. Working in the interpretive zone: Conceptualizing collaboration in qualitative research teams. *Educ. Res.* **1996**, *25*, 5–15.
3. Cummings, J. N.; Kiesler, S. Collaborative research across disciplinary and organizational boundaries. *Social Stud. Sci.* **2005**, *35* (5), 703–722.
4. Steinuehler, C.; Duncan, S. Scientific habits of mind in virtual worlds. *J. Sci. Educ Technol.* **2008**, *17*, 530–543.
5. Kotecha, P.; Walwyn, D.; Pinto, C. *Deepening Research Capacity and Collaboration Across Universities in SADC*; Southern African Regional Universities Association: Johannesburg, 2011.
6. Coburn, C. E. Rethinking scale: Moving beyond numbers to deep and lasting change. *Educ. Res.* **2003**, *32*, 3–12.
7. McDonald, S.; Keesler, V. A.; Kauffman, N. J; Schneider, B. Scaling-up exemplary intervention. *Educ. Res.* **2006**, *35*, 15–24.
8. Holme, T.; Bretz, S. L.; Cooper, M. M.; Lewis, J. E.; Paek, P. L.; Pienta, N.; Stacy, A.; Stevens, R. H.; Towns, M. H. Enhancing the Role of Assessment in Curriculum Reform in Chemistry. *Chem. Educ. Res. Pract.* **2010**, *11*, 92–97.
9. Sandí-Ureña, G. S. *Design and Validation of a Multimethod Assessment of Metacognition and Study of the Effectiveness of Metacognitive Interventions*. Ph.D. Dissertation, Clemson University, Clemson, SC, 2008.
10. Cooper, M. M.; Sandi-Ureña, G. S.; Stevens, R. H. Reliable multi-method assessment of metacognition use in chemistry problem solving. *Chem. Educ. Res. Pract.* **2008**, *9*, 18–24.
11. Xu, X.; Lewis, J. E. Refinement of a chemistry attitude measure for college students. *J. Chem. Educ.* **2011**, *88*, 561–568.
12. Brandriet, A. R.; Xu, X.; Bretz, S. L.; Lewis, J. E. Diagnosing changes in attitude in first-year college chemistry students with a shortened version of Bauer's semantic differential. *Chem. Educ. Res. Pract.* **2011**, *12*, 271–278.
13. Grove, N.; Bretz, S. L. CHEMX: Assessing students' cognitive expectations for learning chemistry. *J. Chem. Educ.* **2007**, *84* (9), 1524–1529.
14. Xu, X.; Lewis, J. E. *Investigating Reliability and Validity of CHEMX Scores from Three Institutions*; 2009, unpublished manuscript.
15. Murphy, K.; Holme, T.; Zenisky, A.; Caruthers, H.; Knaus, K. Building the ACS exams anchoring concept map for undergraduate chemistry. *J. Chem. Educ.* **2012**, *89*, 715–720.
16. Holme, T.; Murphy, K. The ACS Exams Institute Undergraduate Chemistry Anchoring concepts content map I: General chemistry. *J. Chem. Educ.* **2012**, *89*, 721–723.
17. Holme, T.; Murphy, K. Assessing Conceptual versus Algorithmic Knowledge: Are We Engendering New Myths in Chemical Education? In *Investigating Classroom Myths through Research on Teaching and Learning*; Bunce, D. M., Ed.; ACS Symposium Series 1074; American Chemical Society: Washington, DC, 2011; pp 195–206.

18. Villafañe, S. M.; Loertscher, J.; Minderhout, V.; Lewis, J. E. Uncovering students' incorrect ideas about foundational concepts for biochemistry. *Chem. Educ. Res. Pract.* **2011**, *12*, 210–218.
19. McClary, L. M.; Bretz, S. L. Development and assessment of a diagnostic tool to identify organic chemistry students' alternative conceptions related to acid strength. *Int. J. Sci. Educ.* **2010**, *34* (5), 2317–2341.
20. Jiang, B.; Xu, X.; Garcia, C. A.; Lewis, J. E. Comparing two tests of formal reasoning in a college chemistry context. *J. Chem. Educ.* **2010**, *87*, 1430–1437.
21. Beal, C. R.; Stevens, R. H. Improving students' problem solving in a web-based chemistry simulation with embedded metacognitive messages. *Technol., Instruct., Cognit. Learn.* **2011**, *8*, 255–271.
22. Cooper, M.; Stevens, R.; Holme, T. Assessing Problem-Solving Strategies in Chemistry Using the IMMEX system. In *Proceedings of the National STEM Assessment Conference*, National Science Foundation: Washington, DC, 2007; pp 118−129.
23. Bruck, L. B.; Bretz, S. L.; Towns, M. H. Faculty perspectives of undergraduate chemistry laboratory: Goals and obstacles to success. *J. Chem. Educ.* **2010**, *87*, 1416–1424.
24. Bauer, C. F. Attitude towards chemistry: A semantic differential instrument for assessing curriculum impacts. *J. Chem. Educ.* **2008**, *85*, 1440–1445.
25. Cooper, M. M. The case for reform of undergraduate general chemistry curriculum. *J. Chem. Educ.* **2010**, *87*, 231–232.
26. Bretz, S. L. Navigating the landscape of assessment. *J. Chem. Educ.* **2012**, *89*, 689–691.

Chapter 12

Undergraduate Research with Community College Students: Models and Impacts

Thomas B. Higgins*

Department of Physical Sciences, Harold Washington College, 30 E Lake Street, Chicago, Illinois 60601
***E-mail: tbhiggins@ccc.edu**

The role of community colleges in higher education is important and growing, especially with respect to expanding STEM education and undergraduate research opportunities for underrepresented and non-traditional students. The STEM-ENGINES Undergraduate Research Collaborative was an NSF-funded project that provided authentic undergraduate research opportunities for 285 community college undergraduates in the Chicago area. The impacts of this early research experience were increased skill development, enhanced transfer rates, and students' greater realization of their potential to earn STEM degrees. Students from underrepresented groups and first-generation college students were strongly affected, as were their community college faculty mentors.

In 2003, the National Science Foundation announced the Undergraduate Research Centers program, later to be renamed Undergraduate Research Collaboratives (URC). This program was funded by the Division of Chemistry (CHE) and developed in collaboration with the Division of Undergraduate Education (DUE). Susan Hixson was the DUE Program Officer contact for the program (*1*).

The initial URC solicitation was written in response to the workshop report, "Exploring the Concept of Undergraduate Research Centers", which challenged the chemistry community to consider new models of undergraduate research involving collaborative partnerships (*2*). Based on recommendations from this

© 2013 American Chemical Society

report, the goals of the URC program were to (1) expand research opportunities to students in the first and second years of college; (2) broaden participation in undergraduate research, especially with traditionally underrepresented and non-traditional students; and (3) enhance the research infrastructure and change the academic culture of participating institutions by helping them view undergraduate research in a new way.

Later URC solicitations were influenced by the National Academies report "Rising Above the Gathering Storm: Energizing and Employing America for a Brighter Economic Future" (*3*, *4*). A key finding of this report was that, because of increasing globalization, the technical workforce of the United States could no longer depend on attracting the best and brightest scientists from abroad. Therefore, if the US was going to maintain its position as the leading producer of new scientific knowledge and technologies, it would need to increase the number of domestic students pursuing science, technology, engineering, and mathematics (STEM) degrees. To meet the demands of the future STEM workforce, reaching out to underrepresented and non-traditional students was cited as a key step in increasing the quantity of future scientists and engineers. The need to increase the quality of undergraduate science education was also voiced in this report, as well as from other sources (*5*, *6*). Undergraduate research was specifically viewed as a powerful mechanism for recruiting and retaining students STEM students.

Over a three year period, the NSF URC program funded five projects: in 2004, the "Center for Authentic Science Practice in Education" (CASPiE) at Purdue University (*7*); in 2005, the "Ohio Research Experiences to Enhance Learning" (REEL) at The Ohio State University (*8*) and the "Northern Plains Undergraduate Research Center" (NPURC) at the University of South Dakota (*9*); and in 2006, the "Freshman Research Initiative" (FRI) at the University of Texas at Austin (*10*) and the "STEM-ENGINES URC" at Harold Washington College. Each of these projects consisted of a different set of collaborators and a different approach to achieving the core goals of the URC program. Their similarities and differences, as well as an early study of their impacts on students, are described in a 2012 paper published in the Journal of College Science Teaching (*11*). The STEM-ENGINES URC was unique among the five in that it was the only program headquartered at a community college and which had an exclusive focus on community college students.

The Role of Community Colleges in Higher Education

Community colleges play an important role in US higher education and serve a large number of students. During the 2007-2008 academic year, 68% of all undergraduates attended a community college (*12*). Traditionally, the mission of this diverse group of institutions has been to provide convenient and affordable access to higher education for all students, at any stage of life (*13*). They also provide remedial education for students who have completed high school but lack the necessary communication and math skills to pursue college-level courses. Students come to community colleges for a variety of reasons. Although many are looking for a high-quality, affordable first 60 credit hours of a baccalaureate education, a substantial number seek to earn a certificate or associates degree and enter the workforce as soon as possible. Many others are taking a few courses to build essential job skills or for personal enrichment. Among the world's developed countries, there is no other segment of higher education quite like the American community college (*14*).

Community colleges usually draw their funding from a mixture of student tuition (heavily subsidized by state and federal financial aid), local tax revenues, and state funding. With the recent downturn in the economy, however, that funding mix has been shifting away from public revenues and towards student tuition and financial aid.

Some numbers drive home the impact of community colleges on higher education and workforce development: in 2010, almost 13 million students enrolled in one of the nation's 1,132 public, not-for-profit community colleges (*15*). Of this number, 7.68 million students were classified as "credit-seeking" (*16*), meaning they took one or more college-level courses. This represents 42.5% of all credit-seeking undergraduates for that year. For comparison, in 2000, 5.70 million students (48.5% of undergraduates for that year) enrolled at their local community college to earn college credit (*16*).

Because community colleges draw a substantial portion of their funding from local and state coffers, student tuition and fees vary substantially among municipalities and states. On average, annual tuition and fees at a community college are $3,130, less than half of the $8,660 average tuition and fees for a state-run four-year college (*15*). This affordability contributes substantially to the diversity of the community college student body, which spans race, age, financial status, and family history in higher education.

In a given year, community colleges educate almost half of all underrepresented students. Looking at the racial makeup of all US undergraduates in 2011, 49% of all black students, 56% of all Hispanic students, and 42% of all Native American students were enrolled at a community college. The average age of the community college student was 28 (23 was the median) and 15% of all community college students were over the age of 40. During the 2007-2008 academic year, 46% of community college students received some form of financial aid, 34% received Pell grants, and 16% qualified for federal work-study programs. Other important and interesting facts about community college students were that 40% were the first in their family to attend college, 16% were single parents, and 12% were students with disabilities (*15*).

To further complicate the picture, students who take a linear path through college are the exception rather than the rule at community colleges. Undergraduates are much more likely to swirl, bouncing among several institutions of higher education until they earn enough credits to graduate. Often times, students enroll in two institutions at once, swirling between their local community college and a selective-enrollment, baccalaureate-granting institution on a weekly (or daily) basis as they seek to stretch their educational dollars and financial aid as far as possible. This obviously has created complications and has caused many institutions and state legislatures to demand robust articulation and transfer agreements between community colleges and other state-funded institutions so taxpayers are not paying for students to take the same course twice.

With respect to STEM education, a recent National Science Foundation Infobrief reports that over 40% of recent STEM graduates attended a community college during their college career and 28% earned an associate degree prior to transferring to a baccalaureate institution (*17*, *18*). Taken together, these statistics show the significant impact community colleges have on STEM education at all levels of the American educational system. Any effort to increase the quality of STEM education, expand the scientific workforce, or broaden participation in STEM on a large scale must engage community college students and the faculty members who teach them.

Undergraduate Research and Its Impact on Students

Undergraduate research has long been cited as an effective way to attract young minds into the sciences, especially those from underrepresented groups. The American Chemical Society (ACS) has supported early opportunities for economically disadvantaged high school students to pursue research through its Project SEED Summer Research Internship Program (*19*). SEED students not only receive a summer stipend for their time spent doing research, but are also eligible to receive a travel grant to attend an ACS Regional or National Meeting and present their work during their second year of participation. This affords them the opportunity to meet other students like themselves and to interact with professional scientists from academe and industry. Through the ACS Scholars program, the Society supports promising undergraduates from underrepresented groups with generous stipends to support their studies and a mentor to provide advice and guidance (*20*). Both of these programs have proven to be effective methods of recruiting and retaining students into the chemical enterprise.

In addition to being a recruiting tool, undergraduate research has educational benefits for students and produces new knowledge for the discipline. The American Association of Colleges and Universities has cited both undergraduate research and learning communities as two of their five high-impact practices in higher education (*21*). Organizations like the Council on Undergraduate Research (CUR) (*22*) have published numerous examples of the positive impacts of undergraduate research across multiple disciplines in the periodical the CUR Quarterly (*23*) and in the books "Broadening Participation in Undergraduate Research: Fostering Excellence and Enhancing the Impact" (*24*), "Characteristics of Excellence in Undergraduate Research" (*25*), "Developing and Sustaining a Research-Supportive Curriculum: A Compendium of Successful Practices" (*26*), and "Science in Solution: The Impact of Undergraduate Research on Student Learning" (*27*).

Engaging the Next Generation in Exploring Science Undergraduate Research Collaborative: The STEM-ENGINES URC

The STEM-ENGINES URC has tried to take what others have shown is successful in undergraduate research and adapt it to the needs of community college students. The core partners of the collaborative consisted of ten Chicago-area community colleges and three Midwestern baccalaureate-granting institutions. The ten community colleges were the seven City Colleges of Chicago (each separately accredited and strategically distributed throughout the nation's third largest city), William Rainey Harper College (Palatine, IL), Oakton Community College (Des Plaines, IL), and the College of DuPage (Glenn Ellyn, IL). Collectively these community colleges annually served approximately 106,000 undergraduates in their credit division and 42% came from underrepresented groups. At the City Colleges of Chicago, 71% of the 62,000 undergraduates were either black or Latino/a. All of the community colleges were within a two-hour drive of one another, which made periodic meetings as a group relatively easy.

The three four-year partners, by contrast, all were located at least a 2.5 hour drive from Chicago. These institutions were Illinois State University (Normal, IL), Hope College (Holland, MI), and Youngstown State University (Youngstown, OH). This geographic separation was a feature of the initial model of research engagement and will be explained below.

The goals of the collaborative were to develop new models of undergraduate research that incorporated the effective practices of others while they also addressed the needs of our students. Some of the needs observed were (1) being underprepared for college-level work, (2) struggling with economic and financial demands, and (3) having important obligations to their families. Through the success of the STEM-ENGINES URC, there was the potential to not only have a positive impact on students who directly participated, but also be an empowering role model for faculty at the nation's approximately 1,100 community colleges.

Core Ideas, Research Models, and Supporting Activities

The ideas and models of the collaborative had grown out of preliminary work on an NSF Small Grant for Exploratory Research awarded to Harold Washington College and to Harper College, and a classroom research course developed at Oakton Community College. From this initial work, three core ideas had emerged:

(1) Real research questions must be used to challenge students and allow them to fully contribute to the production of new, scientific knowledge;
(2) Scholarly communities of both peers and faculty mentors must support students so they can grow as both scientists and citizens; and
(3) Student transitions beyond the community college, supported by their peers and faculty mentors, are the key to students' long-term academic success and engagement with STEM as a career.

To put these ideas into practice, two models of academic year research at the community college were pursued. This experience was buttressed by summer REU-like research experiences at the four-year college partners and professional development activities designed to build students' skills, nurture their confidence, and expand their network of peer and faculty mentors.

As with any undergraduate research program, students doing research in the laboratory with one or more trusted faculty mentors was the core of the program. At the community colleges, two models of academic year research were pursued (*28*).

(1) A traditional student-mentor model, utilized by faculty mentors at the City Colleges, Harper College, and the College of DuPage. During the academic year, each student pursued his or her own research project, working closely with a single faculty mentor. Students did research in addition to taking a full course load and ideally got credit for their research through an independent study course.

(2) A course-based model, developed and refined at Oakton Community College and later adopted by Harold Washington College. During the academic year, students formally registered for a research course simultaneously taught by up to five faculty members representing multiple disciplines. Course enrollments were low (typically six to ten students, with the course capped at 12 students), which ensured each student received individual attention. At the beginning of the semester, the faculty mentors presented a variety of research options and students collectively decided which to pursue. The students then developed a plan for achieving the class's research goals. These goals were parsed and each student took ownership over a specific subgoal, becoming the local expert and assuming responsibility for instructing his or her peers in that aspect of the project. Many of these projects involved significant interaction with the local community and collaborative projects with institutions like Argonne National Labs and the Chicago Botanical Gardens.

Due to the low student/faculty ratios during the academic year, a close relationship was ensured. From the beginning, students were encouraged to personalize their projects. This increased ownership, engagement, and motivation. All students pursued an authentic project, meaning the answer to the research question and the path to the answer were unknown to everyone. Given that many of the students were beginning their academic careers—some had just enrolled in first semester General Chemistry when their research began—designing such projects was a challenge. But, the authenticity of the research project was of utmost importance because a project for which the results were known was not going to engage students' interests for very long and would actually work against the goal of using research as a method of recruiting promising community college undergraduates into the sciences.

The projects students pursued at the community colleges ran the gamut, from service-learning projects involving the collection and analysis of environmental samples from the local community to benchtop organic synthesis of biologically active small molecules. Although the sizeable majority of the faculty mentors were chemists and the students' projects were chemistry-focused, generous institutional matching funds allowed faculty members from other disciplines such as biology, biotechnology, and geoscience to participate. This multidisciplinary aspect of the collaborative was an unanticipated strength, because the students had the opportunity to see how scientists from different disciplines viewed similar research problems through different lenses. To our delight, almost all our students produced enough results to make presentations at local, regional, and national conferences with an undergraduate focus. Some students produced publications and one was awarded a patent.

Research is a time-intensive endeavor and creating the time for students was essential. Work-life balance is crucial at all stages of a person's career and many students were juggling more responsibilities than the average undergraduate. These responsibilities included caring for parents, children (many of our students were single mothers), and the need to pay for their education and living expenses. To alleviate these pressures, students were paid a modest stipend either from the grant or institutional matching funds. Although the amount of the stipend varied, it was tied to the amount of time the student was able to spend in lab. For example, at the City Colleges of Chicago, students were paid $10 per hour for up to 10 hours a week. To authenticate this, students had to record their hours on a timesheet and the faculty mentor had to verify the time worked. Although this may seem like a modest amount of money, it was important for alleviating students' need to work outside of school.

Students were expected to enroll in a 2-3 credit hour research course whenever possible. The tuition and fees were paid or waived by the institution as part of their support. This class met weekly and provided a venue for all of the students at a college to come together to talk about their research successes and challenges. It also did two other important things: it memorialized the experience on the student's transcript and the credit hours helped the student achieve full-time status. This was important for financial aid eligibility, eligibility for scholarship applications, and other benefits. For example, the City Colleges of Chicago provided all full-time students with a "U-Pass", which allows free public transportation during the academic year. Without the U-Pass, students would pay $86 a month for this transportation benefit.

Supporting the research in the lab were other activities designed to build students' soft skills in areas such as time-management, laboratory safety, and exposure to professional literature. Other foci of the program were building scholarly community, bringing together students, faculty, administrators and visiting scientists from all of the STEM-ENGINES institutions to create an environment where students could learn and talk about science. At the end of each semester, all students presented their research at an in-house poster session, which allowed them to practice their public speaking skills and encouraged them to present their work in public venues such as regional and national meetings.

Each month, the Center for Science Success, a program at Truman College, held professional development seminars for students. This provided a forum in which students could meet their peers and talk about their science without their faculty mentor. At these meetings, students also met with visiting scientists from the community, led discussions on journal articles, and presented their own research roadblocks and accomplishments. Not all mentors required their students to attend, but a majority did.

Working weekends were an opportunity for students to visit a potential summer research site and learn some new chemistry at the same time. These usually lasted a day and a half, including an overnight trip to a new school. The first day usually consisted of a campus visit and visits with faculty from the host institution, followed by a group dinner including students from the host school. The second day included a hands-on workshop during which students learned about modern techniques and instrumentation essential to doing high-quality chemistry.

The working weekends also leveraged the research facilities available at the four-year college partners by sharing them with the less well-equipped community colleges. Since most community colleges relied primarily on local and state taxes for funding, even modest research budgets were not realistic without outside funding. To overcome these difficulties, partnering was essential. Partners provided access to essential resources such as equipment and instrumentation, library collections, and human knowledge and expertise. Thus these partnerships allowed the community college students to access journals and use instrumentation such as NMR and HPLC that were not available without the collaborative.

Following the academic year program, a student would ideally spend seven to ten weeks in residence at one of the four-year college partners. This part of the program was either modeled after the NSF REU program or integrated into an existing REU. Students were expected to devote their full time and attention to research and become part of the research community at a new institution. This was not only an opportunity for students to sample life at another institution, but also a chance to meet a new group of students and faculty and expand their personal scientific network. Importantly, students also had the opportunity to work with dedicated research instrumentation, something the community college budget struggles to maintain and which would not be possible without this sort of partnership.

Although not all students were able to participate in the summer portion of the program, it was a strong belief that this summer at a remote site accelerated their growth as individuals and helped them fully commit to their research (*29*). Promoting personal and professional engagement on a deep level was the reason four-year summer host institutions not in proximity to Chicago were selected. One surprising result was how many students had never traveled far outside of the Chicago-metro area prior to their summer research experience. By pulling students away from their lives in Chicago, they had the opportunity to experience life without the distractions of the city.

Supporting Faculty Mentors

For the faculty mentors, creating the time to mentor students was the most pressing challenge. Traditionally, community colleges have concentrated their resources on providing high-quality, classroom-based instruction. Therefore, faculty members have high teaching loads and may also teach overtime to meet student demand for science courses. At all of the community colleges in this collaborative, 15 contact hours of regular load with an additional 3-6 hours of overtime load per semester was common. In addition, most community colleges relied on a substantial number of adjunct or contingent faculty: up to 50% in some departments. This meant the service component of a faculty member's job had to be distributed among fewer individuals resulting in an increased workload outside of the classroom.

In general, three solutions to these problems were mixed-and-matched as the demands of enrollment dictated:

(1) Faculty were given release or reassigned time, reducing their teaching load. This usually meant a faculty had to mentor a minimum of two students per semester and play an active administrative role on his or her campus with respect to supporting undergraduate research.
(2) The research courses in which students enrolled were counted towards faculty load or overtime.
(3) Faculty were paid overtime in proportion to the number of students mentored.

Although in every case the faculty mentor spent more time working with students than release time was granted, all of the faculty interviewed expressed a high degree of satisfaction with their URC mentoring responsibilities. Indeed, 22 of the 25 faculty members who began the program participated for all five years, a retention rate of 88%.

Student Impacts

Over the five year span of the project, 285 students participated. This headcount is non-duplicated, as many students participated for multiple years. Each year, almost 1/3 of the students returned to the program which created both opportunities and challenges. For example, returning students could be peer mentors and role models for younger students, but also led to the program being oversubscribed, especially in later years.

The student demographics matched reasonably well with the collective demographics of the ten community college in the collaborative and appeared to accurately reflect the diversity of the collective student body of the colleges involved.

Age: 18-63, average 26
Gender: 52% female
Race & Ethnicity:

- 22% Black
- 16% Latino/a
- 20% Asian
- 41% White

One statistic that is particularly pleasing is that no less than 22% of the students were the first generation in their family to go to college. Many also spoke a language other than English in their home or considered English their second language.

The retention and success statistics with the project were impressive. Of the 285 student participants, 274 completed the academic year portion of the project (96%); 135 students did summer research (47%); and 153 students transferred to baccalaureate institutions to complete an undergraduate degree (54%). This last percentage is certainly a low estimate of actual student behavior, for two reasons. First, some of the STEM-ENGINES students are still taking classes at the community college and, second, after students left the project they became harder to track. A longitudinal follow-up to improve the accuracy of this number is currently underway.

Exit interviews with the 11 students who did not complete the project showed that their reasons for leaving varied. Some found the research too rigorous and demanding. Others found it necessary to find higher-paying work. Some had unexpected personal or family issues to deal with. In general, most community college students are juggling multiple responsibilities and important commitments outside of school and this low number was a positive result.

The Survey of Undergraduate Research Experiences (SURE) (*30*) was used to measure the impact the research experience on the students. The SURE measured many things, including increases in students' skills related to research, their plans for academic study beyond the bachelor's degree, and their disciplinary preference. The SURE was administered at the end of the academic year to gauge the impact of that portion of the program. In all three categories, impressive results were witnessed.

In the area of student skills, the SURE instrument asked students to self-report their perceived learning gains in 21 skills germane to becoming an effective practitioner of science. Students ranked these items using a 5-point scale, with 1 = "little or no gain" and 5 = "very large gain". The five skills where students reported the greatest amount of gain were:

- Understanding the Primary Literature (4.4/5.0)
- Understanding Real Problems (4.3/5.0)
- Readiness for More Research (4.3/5.0)
- Feeling Like Part of a Learning Community (4.2/5.0)
- Tolerance for Obstacles (4.2/5.0)

In the area of plans for further academic study, the number of students considering a PhD increased 25 percentage points, from 19% before the academic year research experience to 44% after the experience. When masters degrees were factored in, 63% of our students expressed a desire to earn a degree beyond the bachelors. Correspondingly, the number of students interested in medical school decreased by 8 percentage points. The number not considering graduate school fell 12 percentage points: from 16% to 4% after the academic year program.

Students in the program also reported an increased interest in chemistry as a profession: 37% indicated an interest in majoring in chemistry or biochemistry.

The next most popular choices for a major were biology (28%), engineering (11%), and education (2%).

Another area of success was facilitating and easing student transitions from the community college to the four year college. This work was published in the CUR Quarterly, including three student case studies and lessons we have learned from talking with and listening to the participants (*29*). One of the important lessons learned (and relearned) is the impact of building strong student/faculty connections that span the community college and the baccalaureate institution. These networks support students as they move from the familiar world of the community college to the new and unfamiliar world of a new institution.

Other tangible deliverables include nine peer-reviewed publications in the research literature, eight peer-reviewed publications in the educational literature, and one patent with scientists from Argonne National Labs. Participating students have made over 120 presentations in public forums like the Notre Dame REU Summer Research Symposium, the Chicago Area Undergraduate Research Symposium, the Argonne Symposium for Undergraduates in Science, Engineering, and Mathematics, and Regional and National Meetings of the American Chemical Society. Some students also won prestigious and financially lucrative awards and scholarships such as the American Chemical Society Scholar awards and the Jack Kent Cooke Scholarship.

Based on the student demographics, the retention and success statistics, the SURE data, and the transitions witnessed, it can be stated with confidence that the STEM-ENGINES project met the URC program goals of increasing research opportunities for a diverse student audience and broadening the student talent pool of future STEM majors. Some of the most convincing evidence, though, comes from the students' own words such as this example:

> This research experience changed my life plan completely. *Before this I could not think of myself as going on to attaining my Ph.D. and now I am very confident that I can get that far.* My mentor was a big influence on this decision. He does not take "I can't" for an answer. *(emphasis added)*

Faculty Impacts

Although a thorough study on the impact of research on community college faculty has not been completed, some preliminary interviews have been conducted and they have revealed some important information. First, retention among the faculty mentors was high. Of the 25 community college faculty who participated in the first year, only three did not participate during all five years of the project. This indicates a high degree of satisfaction with the project, despite the fact that mentoring undergraduates generally was more time intensive than classroom teaching.

Some of the impacts faculty have reported during interviews include improving their disciplinary practice, being better able to keep up with emerging topics in their field, increased proficiency with modern instrumentation, expanding their professional networks, better relationships with students, improved teaching, and a greater willingness to innovate in the traditional classroom and laboratory. Expanding these interviews to paint a richer picture of the impact on faculty and how it affects their teaching is in progress. The preliminary work certainly suggests that when a community college faculty member mentors undergraduate research there is a secondary positive impact on the students in the traditional classroom as well.

Conclusions

The STEM-ENGINES URC project affected 285 community college students, affording them opportunities to do undergraduate research at their community college and with partner institutions outside of Chicago. From the assessment data, it is evident that the project achieved the URC goals of expanding research opportunities to students, broadening participation in research, and changing the perception of undergraduate research among community college faculty. This work also provides a positive example of how research impacts community college students for the other 1,100 community colleges in this country. Perhaps someday, programs like this will not be special—they will be ordinary.

Some of the challenges observed while putting these ideas into practice are worth noting, because although significant effort has been expended to address them, they are still vexing issues.

(1) Despite receiving stipends, students still need to supplement their income with other employment. While this is best accomplished with an additional on-campus job, this is not always possible.

(2) Academic year internships with close faculty mentorship are key to helping students grow as researchers and gain the confidence they need to consider leaving home for the summer and pursuing careers in science. Through their research, students become more comfortable and knowledgeable about how science is done and its inherent uncertainties.

(3) Although the project originally envisioned monthly meetings among all URC institutions, in practice the larger community has been strengthened by decreasing the number of large meetings to three times a year. In addition, encouraging smaller groups to meet on their own schedules has created multiple support groups.

(4) Student transitions are best supported with a friendly face and a familiar environment. Willingness of partner faculty members to travel to Chicago and meet with students as well as student travel grants have been crucial to encouraging our students to pursue summer REU opportunities. It has also increased students' awareness of potential transfer institutions.

Acknowledgments

TBH wishes to thank the National Science Foundation for generous funding through two awards from the Division of Chemistry: CHE-0539214 and CHE-

0629174. Morna Rose Brothers, the STEM-ENGINES Project Director, is thanked for her tireless work on behalf of our students, as are all the coPIs and senior personnel of the STEM-ENGINES URC project.

References

1. Undergraduate Research Collaboratives (URC). National Science Foundation, Division of Chemistry. http://www.nsf.gov/funding/pgm_summ.jsp?pims_id=6675 (accessed February 13, 2013).
2. Pemberton, J. E.; Lee, M. Exploring the Concept of Undergraduate Research Centers: A Report on the NSF Workshop. http://urc.arizona.edu/ (accessed February 13, 2013).
3. *Rising Above the Gathering Storm: Energizing and Employing America for a Brighter Economic Future*; National Academy Press: Washington, DC, 2005.
4. Rising Above the Gathering Storm: Energizing and Employing America for a Brighter Economic Future (updated version was released in 2008). http://www.nap.edu/openbook.php?isbn=0309100399 (accessed February 13, 2013).
5. Seymour, E.; Hewitt, N. M. *Talking about Leaving: Why Undergraduates Leave the Sciences*; Westview Press: Boulder, CO, 1997.
6. *How People Learn: Brain, Mind, Experience, and School*; National Academy Press: Washington, DC, 2000.
7. The Center for Authentic Science Practice in Education. Purdue University. http://www.purdue.edu/discoverypark/caspie/ (accessed February 13, 2013).
8. Research Experiences to Enhance Learning. The Ohio State University. http://ohio-reel.osu.edu/ (accessed February 13, 2013).
9. Research and Creative Scholarship Opportunities. The University of South Dakota. http://catalog.usd.edu/content.php?catoid=10&navoid=439#Northern__Plains (accessed February 13, 2013).
10. Freshman Research Initiative. The University of Texas at Austin. https://fri.cns.utexas.edu/ (accessed February 13, 2013).
11. Horsch, E.; St. John, M.; Christensen, R. L. A Case of reform: The undergraduate research collaboratives. *J. Coll. Sci. Teach.* **2012**, *41* (5), 38–43.
12. National Postsecondary Student Aid Study (NPSAS). National Center for Education Statistics. U.S. Department of Education. http://nces.ed.gov/pubs2010/2010205.pdf#page=29 (accessed February 13, 2013).
13. About Community Colleges. American Association of Community Colleges. http://www.aacc.nche.edu/AboutCC/Pages/default.aspx (accessed February 13, 2013).
14. Community Colleges an American Export. National Public Radio. http://www.npr.org/templates/story/story.php?storyId=126646887 (accessed February 13, 2013).

15. AACC Fast Facts. American Association of Community Colleges. http://www.aacc.nche.edu/AboutCC/Pages/fastfactsfactsheet.aspx (accessed February 13, 2013).
16. Advanced Release of Selected 2012 Digest Tables. Table 205. National Center for Education Statistics. http://nces.ed.gov/programs/digest/d12/tables/dt12_205.asp (accessed February 13, 2013).
17. Mooney, G. M.; Foley, D. J. *Community Colleges: Playing an Important Role in the Education of Science, Engineering, and Health Graduates*; NSF 11-317; National Science Foundation, July 2011. http://www.nsf.gov/statistics/infbrief/nsf11317/ (accessed February 13, 2013).
18. Tsapogas, J. *The Role of Community Colleges in the Education of Recent Science and Engineering Graduates*; NSF 04-315; National Science Foundation, April 2004. http://www.nsf.gov/statistics/infbrief/nsf04315/ (accessed February 13, 2013).
19. Project SEED. American Chemical Society. http://www.acs.org/content/acs/en/education/students/highschool/seed.html (accessed February 13, 2013).
20. ACS Scholars Program. American Chemical Society. http://www.acs.org/content/acs/en/funding-and-awards/scholarships/acsscholars.html (accessed February 13, 2013).
21. Brownell, J. E.; Swaner, L. E. *Five High-Impact Practices: Research on Learning Outcomes, Completion, and Quality*; Association of American Colleges and Universities: Washington, DC, 2010.
22. Council on Undergraduate Research. http://www.cur.org/ (accessed February 13, 2013).
23. Council on Undergraduate Research. CUR Quarterly. http://www.cur.org/publications/curquarterly/ (accessed February 13, 2013).
24. Boyd, M. K.; Wesemann, J. L., Eds.; *Broadening Participation in Undergraduate Research: Fostering Excellence and Enhancing the Impact*; Council on Undergraduate Research: Washington, DC, 2009.
25. Hensel, N., Ed.; *Characteristics of Excellence in Undergraduate Research*; Council on Undergraduate Research: Washington, DC, 2012. http://www.cur.org/publications/publication_listings/COEUR/ (accessed February 13, 2013).
26. Karukstis, K. K.; Elgren, T. E. *Developing and Sustaining a Research-Supportive Curriculum: A Compendium of Successful Practices*; Council on Undergraduate Research: Washington, DC, 2007.
27. Lopatto, D. *Science in Solution: The Impact of Undergraduate Research on Student Learning*; Research Corporation for Science Advancement: Tucson, AZ, 2009.
28. Higgins, T. B.; Brothers, M. Research models that engage community college students. *CUR Quarterly* **2008** , Fall, *29* (1), 12–17.
29. Higgins, T. B.; Brown, K. L.; Gillmore, J. G.; Johnson, J. B.; Peaslee, G. F.; Stanford, D. J. Successful transitions from the two-year college to the four-year college facilitated by undergraduate research. *CUR Quarterly* **2011** , Spring, *31* (3), 16–22.
30. Lopatto, D. Survey of undergraduate research experiences (SURE): First findings. *Cell Biol. Educ.* **2004**, *3*, 270.

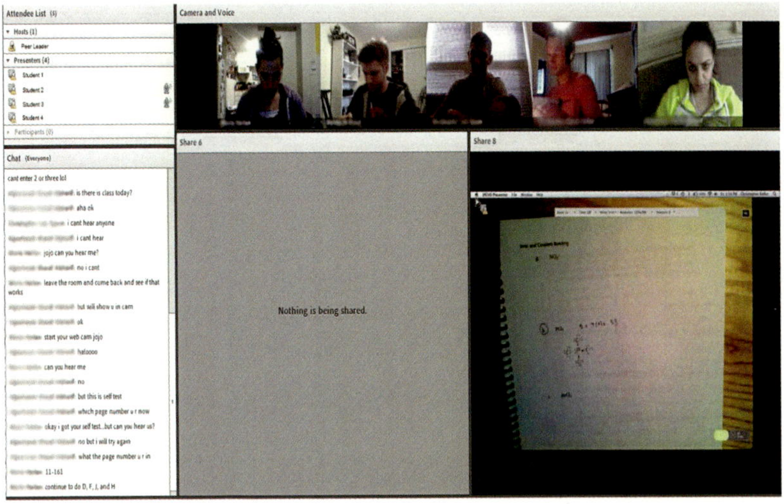

Figure 1. Basic structure of a cPLTL workshop. (Chapter 7)

Figure 2. cPLTL technology set. (Chapter 7)

Chapter 13

Preparing the Future STEM Faculty: The Center for the Integration of Research, Teaching, and Learning

Robert D. Mathieu*

Department of Astronomy, University of Wisconsin–Madison, Madison Wisconsin 53726
***E-mail: mathieu@astro.wisc.edu**

The Center for the Integration of Research, Teaching, and Learning (CIRTL) seeks to enhance excellence in under-graduate education through development of a national STEM faculty committed to implementing and advancing effective teaching practices for diverse learners. Graduate education is a powerful leverage point to develop such a national faculty; nearly 80% of all U.S. Ph.D.'s are trained at only 100 research universities. CIRTL has developed, implemented, and evaluated strategies that prepare future faculty for careers that integrate research, teaching, and learning based on three core ideas: teaching-as-research, learning communities, and learning-through-diversity. Evaluation shows that the learning outcomes of high-engagement CIRTL participants reflect research-based, high-impact teaching practices, while longitudinal studies indicate that they use the knowledge and skills they gained from teaching development in subsequent undergraduate teaching. Currently the CIRTL Network comprises 22 major research universities.

Introduction

The prosperity of U.S. society in a global economy depends upon a competitive, technically expert, college-educated workforce (*1*). Yet there continue to be serious concerns about the quality of STEM undergraduate education that relate to the uneven quality of teaching (*2–4*), low enrollment

© 2013 American Chemical Society

and retention in introductory courses and inadequate student understanding (*5–7*), and the relationship between classroom experiences and equitable student achievement and persistence in STEM (*8–11*). In the face of these ongoing concerns, there is a critical need to prepare future faculty members to be both leading researchers *and* excellent teachers.

The *Center for the Integration of Research, Teaching, and Learning* (CIRTL) was founded as an NSF Center for Learning and Teaching in higher education. **CIRTL seeks to enhance excellence in undergraduate education through the development of a national faculty committed to implementing and advancing effective teaching practices for diverse learners as part of successful and varied professional careers.** The goal of CIRTL is to help improve the STEM learning of all students across the higher education landscape, and thereby to increase the diversity in STEM fields and the STEM literacy of the nation.

The strategic leverage point through which CIRTL seeks to help shape the future of STEM undergraduate education is graduate education at research universities. Nearly 80% of STEM PhDs are granted at only 100 research universities (*12*), allowing for a highly targeted intervention before graduates flow into faculty positions at the 4400 U.S. research universities, comprehensive universities, liberal arts colleges, and community and tribal colleges.

CIRTL has developed, implemented, and evaluated strategies for preparing STEM future faculty (*13*) for careers that integrate research, teaching, and learning based on three core ideas: **teaching-as-research**, **learning communities**, and *learning-through-diversity*. CIRTL established a prototype learning community at the University of Wisconsin–Madison (UW) in 2003 (*14*, *15*); demonstrated that the CIRTL core ideas can be integrated into an existing graduate professional development program at Michigan State University (MSU) in 2005; and launched a successful prototype CIRTL Network of six diverse universities in 2007: University of Colorado at Boulder, Howard University, MSU, Texas A&M University, Vanderbilt University, and UW. Currently the CIRTL Network comprises 22 major research universities.

Here we present an overview of the three core ideas, provide an overview of the design of the UW CIRTL learning community, present evaluation findings on the learning outcomes of CIRTL participants, describe the national CIRTL Network and provide several closing thoughts.

The CIRTL Core Ideas

CIRTL's conceptual framework consists of three core ideas: teaching-as-research, learning community, and learning-through-diversity. Specifically:

Teaching-as-research (TAR) is the deliberate, systematic, and reflective use of research methods by STEM instructors to develop and implement teaching practices that advance the learning experiences and outcomes of both students and teachers.

Learning communities (LC) bring together groups of people for shared learning, discovery and generation of knowledge. To achieve common learning goals, a learning community nurtures functional relationships among its members.

Learning communities serve as both formal and informal spaces that encourage and support transformation of the teaching–learning process.

Learning-through-diversity (LtD) recognizes that excellence and diversity are necessarily intertwined. True learning-through-diversity capitalizes on the rich array of experiences, backgrounds, and skills among STEM undergraduates and graduates-through-faculty to enhance the learning of all.

These ideas help CIRTL participants build parallels between their approaches to research and to teaching, thereby lower barriers to engaging in advancing learning. Teaching-as-research describes the process of improving student learning in terms that are familiar from disciplinary research, while the core ideas of learning community and learning-through-diversity emphasize the rich and productive experiences researchers have working in diverse teams to achieve a common goal. These ideas are rooted in the practice of STEM research teams, the educational literature, and past initiatives in faculty development such as the Preparing Future Faculty program (*16*). CIRTL places these three core ideas at the very foundation of preparation for STEM teaching and learning.

These ideas also operate powerfully at multiple levels. First, they lie at the heart of the learning objectives of CIRTL professional development activities. Each activity seeks to enable graduates-through-faculty (*17*) —throughout their careers—to create learning communities of their students, to practice teaching-as-research, and enhance the learning of all students. Second, the graduates-through-faculty themselves form campus learning communities that enable members to investigate the effects of teaching practice and capitalize on their diverse perspectives. Third, the cross-Network learning community enables all Network future faculty to learn from the diversity of graduate-through-faculty experiences, university cultures, etc. of the CIRTL Network. Finally, the leaders and implementers of the CIRTL campus learning communities themselves form a Network learning community sharing assessments and outcomes, resources, experiences, and ideas with each other and the nation.

CIRTL has established a detailed set of learning outcomes for Network future faculty (www.cirtl.net/CIRTLoutcomes). Achievement of the outcomes is organized in three developmental levels. The CIRTL **Fellow** is able to implement research-based best practices to achieve defined learning goals. The CIRTL **Practitioner** practices scholarly teaching that uses the CIRTL core ideas to demonstrably improve learning. The CIRTL **Scholar** produces public scholarship that advances teaching and learning. CIRTL outcomes conceived in this way permit anyone to enter a CIRTL Network learning community from a wide variety of disciplines, needs, and past experiences, and to advance their abilities as a teacher at measurable achievement levels.

The Delta Program in Research, Teaching, and Learning

CIRTL's first major goal was to develop, implement, evaluate, and institutionalize an effective STEM graduate-through-faculty learning community, centered on preparing future faculty in teaching and learning and founded on the CIRTL core ideas. The initial CIRTL partners – Michigan State University, the Pennsylvania State University, and UW – chose to use UW as the primary laboratory for design, implementation, evaluation and research. This prototype CIRTL learning community is called the *Delta Program in Research, Teaching, and Learning* (www.delta.wisc.edu).

The programmatic component of Delta comprises graduate courses, small-group facilitated programs for graduates-through-faculty, mentor training, TAR internships, and ancillary workshops. The program design emphasizes semester-long intervals of engagement, building on research showing that such longer term engagement is more transformational (*18*). Every facet of Delta is designed around research models familiar to STEM graduates-through-faculty. The courses are project-based, requiring graduate students to define a learning problem; understand the undergraduate audience; explore the literature for prior knowledge; hypothesize, design, and implement a solution; and acquire and analyze data to measure learning outcomes. The Delta internships are research assistantships in teaching, in which a graduate student or post-doc partners with a faculty member to address a learning problem. The Delta activities are designed to provide each graduate and post-doctoral participant with a portfolio, letters of recommendation, and presentations/publications in teaching and learning analogous to those in their disciplinary research curriculum vitae. And finally, courses are often team-taught by research-active STEM and social science faculty and staff. These pairings provide powerful combinations of experience, theoretical foundation, and role modeling for the STEM future faculty.

The Delta Program, launched in 2003, has included 2,380 STEM graduate-through-faculty participants through 2012. 40% of the future faculty engage in 15 or more hours of Delta programs. The participants comprise 21% physical and mathematical sciences, 45% biological sciences, 18% engineering sciences, and 13% social, behavioral, and economic sciences. The gender distribution among graduate students is nearly equal.

All standard operations of the Delta Program have been fully funded by UW since 2007, so this prototype CIRTL learning community is successfully institutionalized. External funding continues to be obtained for further development of the Delta programming and learing community.

CIRTL Learning Outcomes in the Delta Program

Between fall 2005 and fall 2009 312 high-engagement participants (greater than 15 contact hours) in 39 offerings of the Delta Program answered two questions upon completion of a program:

What major concepts are you taking away from this Delta course, program, or activity that will affect your practice as an educator? If possible, please give two to three specific examples.

Suppose that you are preparing to teach some scientific concept from your discipline (e.g., the nitrogen cycle, amplitude, redox reactions). Describe the steps that you will take, based on what you've learned in this course, program, or activity.

Figure 1 shows the major concepts and approaches to teaching presented by respondents (*15*). The learning outcome categories were derived from their responses, not from preconceived notions of what they "should have" learned through their participation in Delta. Just over 90% expressed TAR ideas; for example 74% discussed *assessment/evaluation* and half explicitly called out defining *learning outcomes*. 57% integrated the presence of diverse learners into their thinking about teaching, including concepts such as *inclusive teaching* and *diverse instruction*. Nearly half included learning community ideas in their responses, and especially *group work*. Additional learning outcomes were expressed but did not fit neatly within one of the CIRTL core ideas. For example, 72% noted the importance of understanding *learners and learning*, with particular emphasis on cognition, learning and development, and knowing students' backgrounds and perspectives.

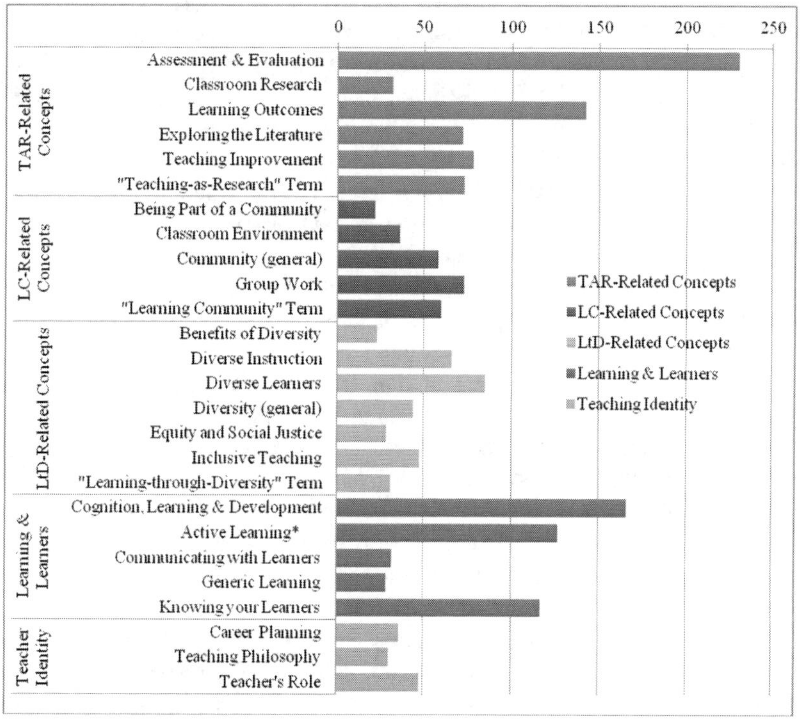

Figure 1. Learning outcomes of CIRTL future faculty participating in high-engagement activities. The listed items to the left of the bars are the data categories of concepts and approaches to teaching that are summed together in each bar. Responses are organized into subcategories within each major conceptual category. N = 312 respondents; overall response rate was 68%. The responses came from a wide range of participants with different career stages, varied levels of participation in the Delta Program, and diverse disciplines. (Reproduced with permission from reference (15). Copyright 2012 Taylor & Francis).

Current education research (*19–23*) supports the argument that the national goal of advancing STEM undergraduate learning will be advanced by STEM faculty who characterize and engage in their teaching similarly to the future faculty in this study. The CIRTL hypothesis has been that future faculty will embrace these research-based, high-impact approaches to teaching by doing teaching-as-research, having learning community experiences, and experiencing learning-through-diversity—and furthermore, that their *self*-discovery will lead to deeper understanding and engagement with these ideas. This hypothesis aligns with current understanding of STEM undergraduate learning, which emphasizes the importance of student engagement with STEM ideas. These data support this hypothesis, although attribution of outcomes is always difficult.

In 2005 CIRTL initiated a longitudinal study of 83 future faculty (*24*). Of the 67 still responding in 2011, 80% remain in higher education, 49% are currently associated with undergraduate education, and 30% are in tenure-track faculty positions. For the last group, half are in predominantly undergraduate institutions.

Respondents' *current* perceptions of learning gains from their CIRTL experiences fit into four broad thematic categories: *diversity of perspectives* (e.g., the most commonly reported cognitive gains related to diversity in the classroom); *importance of engaging students in active learning*; *connections between teaching and scientific research* (e.g., that the teaching process can be enhanced by scientific methods); and *design and organization to meet specific learning goals*.

A majority of study respondents (76%) found ways to use the knowledge and skills they gained from teaching development in their subsequent undergraduate teaching. Respondents most frequently cited delivering *instruction that increases student engagement* (e.g., through active learning techniques, inquiry-based learning, or the creation of learning communities within the classroom). They also frequently cited what they had learned in *assessment* and *course preparation and planning*, especially backward design by starting with learning goals.

Respondents reported that their participation contributed significantly to their early-career success, namely: job satisfaction, peer approbation, and membership in learning communities. When asked what influenced their effectiveness in their current job, the respondents linked their effectiveness to how well they thought their students were performing in class and how well they felt they themselves were balancing work and personal responsibilities. When asked how they felt their colleagues, peers, students, or supervisors gauged their effectiveness, nearly 90% responded positively, based on formal and informal processes.

This prototype CIRTL learning community demonstrated that a major research university can and will commit to the preparation of STEM graduate students to be both forefront researchers and excellent teachers. In addition, the Delta experiment established that there is a strong felt need for such preparation among STEM graduate students and postdocs, and a willingness of STEM faculty to provide and support that preparation. Finally, Delta demonstrated that a graduate-through-faculty learning community built on the CIRTL pillars is an effective approach at a research university both to improve preparation for teaching and to promote institutional change.

Furthermore, starting in 2005 MSU demonstrated that the CIRTL core ideas can be integrated into an existing graduate professional development. The MSU *Future Academic Scholars in Teaching (FAST) Fellowship Program* is a TAR-based learning community for STEM doctoral students (*25*). Now in its sixth year, it is oversubscribed by a factor three. FAST Fellows have participated from 20 departments. To broaden impact, MSU offers a 1.5-day teaching and learning institute to introduce future faculty to high-impact practices, learning-through-diversity, and TAR, and to help them plan TAR projects. These programs, and many CIRTL-based workshops, are institutionalized within the broader MSU *PREP* professional development program.

The Prototype CIRTL Network

To prepare the future STEM faculty of the nation, CIRTL seeks to similarly influence future faculty preparation in teaching and learning at a significant number of research universities. Building again on the CIRTL pillars, we have developed the *CIRTL Network*, a learning community of diverse research universities mutually engaged in TAR activities to prepare future faculty in teaching and learning for all students.

To test this strategy, CIRTL first created a small prototype CIRTL Network. Established in Fall 2006, the prototype CIRTL Network comprised the University of Colorado at Boulder, Howard University, Michigan State University, Texas A&M University, Vanderbilt University, and the University of Wisconsin - Madison. The diversity of these institutions—private/public; large/moderate size; majority-/minority-serving; geographic location—was by design.

Critically, the CIRTL Network is a learning community rather than a confederation. As such, the CIRTL Network is a path for mutual adaptation, participation, and collaboration. Every university provides a different context; indeed, the diversity of institutions, programs, and people is a primary motivation for the CIRTL Network. **A key hypothesis is that the preparation in teaching and learning of a graduate student or postdoc at any CIRTL Network university will be substantially enhanced, directly and indirectly, as a result of the diversity across the Network.**

The CIRTL Network enhances the preparation in teaching and learning of STEM future faculty in at least three ways: (a) through the development and enhancement of local CIRTL learning communities, building on successes throughout the Network; (b) through cross-Network programs that expand the graduate-through-faculty learning community beyond the local university; and (c) through the development of a community that extends beyond graduate school into the faculty experience. Here we focus on the cross-Network learning community.

As of 2012, the cross-Network learning community regularly offered these online synchronous opportunities: four or five semester-long CIRTL courses; the *CIRTL Capstone TAR Seminar*, forming a learning community of TAR interns across the Network; *CIRTL Coffee Hours*, a series for future faculty to connect informally in discussions of topics of mutual interest; and occasional *CIRTLCasts*, webinars by guest experts. Capstone experiences for TAR interns include *CIRTL Exchanges* which send future faculty to Network campuses to present both disciplinary research and TAR seminars. These exchanges are valuable both for the TAR intern *and* for promoting CIRTL on the host campus, as the future faculty themselves are invariably the best ambassadors.

Since 2008 participation in the cross-Network learning community has increased by an average of a factor of three each year, with 834 participations in 2012. Currently participation is limited only by capacity, with waitlists each semester.

When reflecting on the impact of a cross-Network learning community, participants most often cited the diversity of ideas and experiences among classmates and instructors (*26*). Participants indicated that the courses exposed them to a diversity of ideas with which they were unfamiliar, both through the content of the courses and the diverse institutional contexts of participants. Participants indicated that the diversity of institutional representation enhanced the feedback on their course or TAR projects due to different perspectives on teaching, and broadened their understanding about how other institutions organized and managed their courses.

Our initial approach to the cross-Network learning community emphasized synchronous interaction, reflecting the high priority of learning-through-diversity across the Network. At the same time, CIRTL has invested substantially in a web-based portal (www.cirtl.net) that facilitates Network-wide engagement, collaboration, and participation in online programming. We have begun development of a set of four asynchronous learning community sites organized around themed areas of intellectual content central to CIRTL's goals. By connecting the themed areas with CIRTL Network courses, the sites will gain a regular influx of new people and contemporary resources, while providing a long-term presence that brings together learning community members.

Expansion of the CIRTL Network

Based on the success of the prototype network, in 2011 the CIRTL Network went through a deliberate process for a major expansion to more than 20 universities. This expansion of the CIRTL Network was a carefully designed and executed process, with two primary goals: (1) bring in institutions with substantial impact on the national STEM faculty through placement of large numbers of graduates in undergraduate faculties or through other significant impact; and (2) further expand the diversity of institutions and expertise available to the Network learning community.

In 2010 the six prototype network universities developed membership, operations, and financial plans and an organizational structure, all presented in *CIRTL for the Nation: A Growth Plan* (www.cirtl.net/expansion/growthplan). In March 2011 a call for applications was sent to the provosts and graduate deans of the top 100 STEM Ph.D. producers in the United States, as well as to universities requesting invitations based on earlier outreach. Thirty-five universities submitted applications, and in June 2011, offers were made. As of March 2012, the CIRTL

Network comprises 22 major universities (Table 1). These universities represent 22% of the Ph.D. productions of the United States.

Table 1. The CIRTL Network – 2013

Boston University	University of Alabama at Birmingham	University of Pittsburgh
Cornell University	University of California, San Diego	University of Rochester
Howard University	University of Colorado at Boulder	University of Texas at Arlington
Iowa State University	University of Georgia	University of Wisconsin–Madison
Johns Hopkins University	University of Houston	Vanderbilt University
Michigan State University	University of Maryland, College Park	Washington University in St. Louis
Northwestern University	University of Massachusetts Amherst	
Texas A&M University	University of Missouri–Columbia	

Closing Thoughts

The National Science Foundation, as well as many other agencies and foundations, and innumerable people across the nation, has invested heavily in determining through research and evaluated practice what are high-impact teaching practices that enhance STEM undergraduate learning (e.g, (*19*)). Even so, research studies repeatedly find that the outcomes of this investment are far too rarely implemented in college and university teaching (2). Arguably, strategic directions for improvement of undergraduate success and retention, including reducing the achievement gap, today lie less in the development of new ideas and more in the broad implementation of known research-based high-impact teaching practices.

For many reasons, the nation's future STEM faculty represent a powerful and accessible leverage point for achieving this strategic goal. First, graduate students and post-doctoral fellows are highly concentrated in approximately 100 universities, allowing a highly targeted intervention before those who will become faculty disperse along varied paths into more than 4000 institutions of higher education. Second, future faculty are not yet set in their perspectives on ways of teaching. Third, having recently been STEM undergraduates, we find that graduate students resonate with the need for reform in STEM undergraduate education, and on behalf of their future teaching specifically. And finally, perhaps more than many research university faculty, graduate students and post-docs appreciate the importance of being both excellent researchers and excellent teachers in order both to get jobs and to have successful careers within the diverse landscape of higher education.

Thus it is perhaps no surprise that a universal experience across the CIRTL Network is that the future faculty demand for preparation in teaching is very strong, typically exceeding capacity. Key tactical issues then become adequate investment of research universities in future faculty preparation, and permission, explicit or implicit, of research advisors for future faculty to participate. In truth, achievement of the CIRTL mission is as much about shifting research university cultures as about the processes of preparing future faculty in teaching. Such cultural change is one primary motivation behind the ideas of teaching-as-research and learning communities.

The success of the TAR idea in lowering cultural barriers and initiating a cultural shift lies in its alignment with the current skills and beliefs of STEM faculty. TAR places engagement in advancing teaching in alignment with engagement in STEM research. The improvement of teaching is itself a research problem, addressing the question "What have my students learned?". Foundational knowledge from disciplinary literature, hypotheses and goals, experimental constructs, collecting and analyzing data from the classroom are approaches and words that align advancing student learning with STEM research. This alignment can be carried further into implementations with designs – such as teaching-as-research assistants - that parallel research models familiar to STEM graduates-through-faculty.

A powerful concept requires a community in which to develop and flourish if broad change is to result. Thus CIRTL creates interdisciplinary learning communities around teaching-as-research which, critically, are of and by STEM graduates-through-faculty. Learning communities are enduring and integrative environments for change in teaching and learning. At the same time, learning communities also foster strong relationships among members across an institution and thus build a foundation for institutional change. These core ideas of CIRTL engage STEM graduates-through-faculty in the cultural change of future faculty preparation – they do not sit outside the change. Ultimately it is this broad interdisciplinary engagement that shifts norms and yields institutionalization in the broadest sense.

The CIRTL focus on graduate education is situated within a systems perspective that recognizes the complex set of factors impacting student learning. No single project alone can transform undergraduate learning. The CIRTL Network stands in partnership with a strong array of national and local initiatives seeking transformational change in the skills, attitudes, and norms of both future and current faculty with respect to STEM undergraduate teaching and learning. Importantly, we envision—and already see—that the future faculty of CIRTL and the participants of other national initiatives in teaching and learning will become partners and leaders of current and future movements for change.

Acknowledgments

Many hundreds of graduate students, post-docs, faculty, and academic staff across the natural and social sciences have contributed to the success of CIRTL participants in learning communities across the CIRTL Network. Still, I would be remiss in not acknowledging two members of CIRTL leadership who have guided CIRTL for more than a decade: Ann Austin, professor of higher education at Michigan State University and Katherine Barnicle, the executive director of CIRTL. Without their wisdom and commitment CIRTL as it stands today would not have happened.

The same is true of the ongoing support of the National Science Foundation, through the funding of grants DRL-0227592 and DUE-0717768.

No less important were the contributions of Dr. Susan Hixson, the NSF program officer for CIRTL over the decade. She facilitated this large and complex program administratively at NSF, contributed intellectually at each meeting of the National Advisory Board, and made connections for CIRTL across the nation. The successes of CIRTL are in no small part the result of Susan's wise and committed guidance, which all of us in CIRTL honor here with respect, appreciation and deep gratitude.

References

1. National Research Council. *Rising Above the Gathering Storm, Revisited: Rapidly Approaching Category 5*; National Academies Press: Washington, DC, 2010.
2. Fairweather, J. *Evidence of Promising Practices in Undergraduate STEM Education*; Commissioned paper for the National Academies Workshop; 2008.
3. Seymour, E.; Hewitt, N. M. *Talking about Leaving: Why Undergraduates Leave the Sciences*; Westview Press: Boulder, CO, 1997.
4. Stage, F. K.; Kinzie, J. *J. Gen. Educ.* **2009**, *58* (2), 85–105.
5. National Research Council. *Evaluating and Improving Undergraduate Teaching in Science, Technology, Engineering, and Mathematics*; National Academies Press: Washington, DC, 2003.
6. Project Kaleidoscope. Transforming America's Scientific and Technological infrastructure: Recommendations for Urgent Action; Report on Reports II; 2006. http://www.nigms.nih.gov/NR/rdonlyres/90AA6B67-2945-4923-BBE4-7E33D9505469/0/2006ReportOnReports.pdf.
7. Shaw, E. J.; Barbuti, S. *NACADA J.* **2010**, *30* (2), 19–34.
8. Cabrera, A. F.; Colbeck, C. L.; Terenzini, P. T. *Res. Higher Educ.* **2001**, *42*, 327–352.
9. Espinosa, L. L. Ph.D. Thesis, University of California, Los Angeles, CA, 2009.
10. National Research Council. *Expanding Underrepresented Minority Participation: America's Science and Technology Talent at the Crossroads*; National Academies Press: Washington, DC, 2010.

11. Sonnert, G.; Fox, M. F.; Adkins, K. *Social Science Quarterly* **2007**, *88*, 1333–1356.
12. National Science Foundation, Division of Science Resource Statistics. Doctorate Recipients from U.S. Universities: Summary Report 2007–08; NSF 10-309; 2009. http://www.nsf.gov/statistics/nsf10309/.
13. In this article, "faculty" includes all undergraduate instructors, and "future faculty" means graduate students and post-doctoral fellows, recognizing that all may not follow career paths that include undergraduate education.
14. Mathieu, R. D. *Teaching as Research: A Concept for Change at Research Universities*. Paper presented at the International Colloquium on Research and Teaching: Closing the Divide, Winchester, England, 2004. http://www.cirtl.net/files/MathieuTeaching-as-ResearchFeb2004.pdf.
15. Pfund, C.; Mathieu, R.; Austin, A. E.; Connolly, M.; Manske, B.; Moore, K. *Change* **2012**, November/December.
16. Pruitt-Logan, A. S.; Gaff, J. G. In *Paths to the Professoriate: Strategies for Enriching the Preparation of Future Faculty*; Wulff, D. H., Austin, A. E., Eds.; Jossey-Bass: San Francisco, CA, 2004; pp 177−193.
17. Graduate students, post-doctoral fellows, academic staff and faculty.
18. Garet, M. S.; Birman, B. F.; Porter, A. C.; Desimone, L.; Herman, R.; Yoon, K. S. *Designing Effective Professional Development: Lessons from the Eisenhower Program*; U.S. Department of Education, Office of the Under Secretary: Washington, DC, 1999. http://eric.ed.gov/?id=ED442634(accessed January 2, 2007).
19. National Research Council. *Discipline-Based Education Research: Understanding and Improving Learning in Science and Engineering*; The National Academies Press: Washington, DC, 2012.
20. National Research Council. *How People Learn*; The National Academies Press: Washington, DC, 2000.
21. Ambrose, S. A.; Bridges, M. W.; Lovetts, M. C.; DiPietro, M.; Norman, M. K. *How Learning Works: Seven Research-Based Principles for Smart Teaching*; Jossey-Bass: San Francisco, CA, 2010.
22. Svinicki, M.; McKeachie, W. *McKeachie's Teaching Tips: Strategies, Research, and Theory for College and University Teachers*, 13th ed.; Houghton Mifflin Company: Boston, 2010.
23. Weimer, M. *Learner-Centered Teaching: Five Key Changes to Practice*; Jossey-Bass: San Francisco, CA, 2002.
24. Benbow, R.; Byrd, D.; Connolly, M. *The Wisconsin Longitudinal Study of Doctoral and Postdoctoral Teaching Development: Key Findings*; University of Wisconsin–Madison, Center for the Integration of Research, Teaching, and Learning: Madison, WI, 2011.
25. Vergara, C. E.; Campa, H., III; Cheruvelil, K.; Ebert-May, D.; Fata-Hartley, C.; Johnston, K.; Urban-Lurain, M. *Innovative Higher Educ.* **2013**, *39* (2).
26. Shelton, L. J.; Micomonaco, J. *Evaluation Findings of CIRTL Cross-Network Courses 2009−2011*; CIRTL: Michigan State University: East Lansing, MI, 2012.

Chapter 14

Improving STEM Student Success and Beyond: One STEP at a Time

Maureen A. Scharberg*

Student Academic Success Services, Clark Hall 105, San José State University, San José, California 95192-0018
***E-mail: Maureen.Scharberg@sjsu.edu**

Six years ago, San José State University (SJSU) received from the National Science Foundation a STEM Talent Expansion Program (STEP) grant (Type 1A, Grant #0653260). We embarked on a journey to transform our STEM student culture by implementing a comprehensive support program. Our College of Science STEP program has the following elements: mandatory academic advising, progress to degree program, College of Science Advising Center (COSAC), supplemental instruction, a probation course, and peer advising and tutoring. In the final year of our project, we have had a measurable increase in the retention of STEM majors. Many of these key elements have been institutionalized. Four other SJSU colleges have started their own student success centers, following the COSAC model, with support from the Provost's Office. The overall effect of our STEP grant along with Susan's support has definitely transformed the student success culture at SJSU.

© 2013 American Chemical Society

Introduction

The College of Science at San José State University was awarded from the National Science Foundation (NSF) a STEM Talent Expansion Program (STEP) grant in 2007, entitled "Improving Retention Through Student Learning Communities". The NSF STEP grant program seeks to increase the number of associate and bachelor STEM degrees awarded to students. The objectives of our STEP grant were to:

1. Expand and enhance academic and career advising to entering students,
2. Provide professional development opportunities for faculty who teach STEM "gateway" courses, and
3. Immerse STEM majors into comprehensive learning communities.

Much of the development of this grant was based on lessons learned in the "New Traditions" Chemistry Project that was one of the five projects funded through the NSF Initiative Systemic Changes in the Undergraduate Chemistry Curriculum (*1–3*). Our STEP grant focused on student-centered, active learning communities, faculty professional development that was discipline specific and the use of data to drive curricular improvement and enhancements. The trajectory for this project extended beyond the chemistry discipline reaching out to the entire campus to improve undergraduate retention and graduation rates at San José State University, after focusing initially on other STEM fields (computer science, math and physics). Our chemistry department had already achieved some of the goals of this proposal, so our STEP grant, being a college-based proposal, chose to focus on these three STEM disciplines.

San José State University (SJSU), located in San José, California, the heart of Silicon Valley, is the oldest campus in the California State University (CSU) system. SJSU is a fully-accredited, public, comprehensive university offering bachelor's and master's degrees in 134 areas of study. SJSU offers rigorous course work to more than 30,000 undergraduate and graduate students in seven colleges. As one of the 23 campuses in the CSU system, SJSU is a leader in high-quality, accessible, student-focused higher education. The College of Science is home to approximately 2000 undergraduates and offers undergraduate degree programs in biological sciences, chemistry, computer science, geology, mathematics, meteorology & climate science as well as physics & astronomy.

Before the STEP grant, the College of Science was struggling with student success for not only our majors but also for engineering students who enrolled in several of our science gateway courses. After analyzing the number of repeats in STEM courses such as calculus, chemistry and physics, it was clear that students were not making progress to degree in a timely fashion. Some students enrolled in pre-calculus and calculus courses more than twice before passing these courses with a final course grade of "C" or better. Others would try a gateway STEM course and not pass. Then, they would focus on completing non-STEM lower division general education courses and not repeat STEM gateway courses. Thus, we realized that STEM student success was complicated and would need to involve several types of interventions for both our first year student cohorts and also our transfer student cohorts (*4–10*).

From analyzing transcripts, faculty advisors had observed that students who struggled through our gateway courses in calculus, chemistry and physics would wind up on university probation and even disqualified from SJSU. We also inferred that struggling students needed more intrusive academic advising to monitor progress to degree, including keeping them off of probation and repeating courses. Intrusive academic advising is proactive academic advising, geared to regularly assessing students' timely progress to degree. Timely progress to degree is defined as within four to six years for full-time first year students. For example, in the Colleges of Science and Engineering at SJSU, STEM undergraduates cannot register for classes for the upcoming semester unless they meet with either a staff or faculty academic advisor. A hold is placed on their registration and is lifted after their academic advising appointment.

Given the student outcomes from the NSF "New Traditions" project, we realized the potential to increase students success by transitioning to student-centered, research-based pedagogy in our gateway courses. Our new students needed to successfully transition to the College of Science and SJSU and develop a sense of belonging to these communities. We also wanted to give our students leadership opportunities through peer advising, so they could help mentor new students and tutor students who might need extra assistance with their STEM coursework.

This paper highlights the efforts in this process associated specifically with our STEP grant and describes our lessons learned, according to the goals of our grant. We also discuss strategies for sustaining such programs and show how other colleges at SJSU benefited from the outcomes from our STEP grant. Although this grant is outside the typical chemistry education research framework, the information learned from our STEP efforts are applicable to any STEM department.

Expanding and Enhancing Academic and Career Advising to Entering Students

Given that the College of Science's departments were located in three buildings, we felt that our students needed a "one-stop shop" where they could find both academic and career advising. In April 2008, the College of Science Advising Center (COSAC) opened to offer these services to students. COSAC is staffed with three academic advisors who provide lower division major academic advising for Biological Sciences, Chemistry, and Computer Science in partnership with faculty advisors in these departments. Because the other College of Science departments are smaller, faculty advisors in these departments provide all major advising to their undergraduates. COSAC advisors can also assist all College of Science undergraduates with general College of Science advising, General Education inquiries and transfer articulation issues. An office manager oversees the day-to-day operations and supervises the student peer advisors who are located in COSAC.

COSAC peer advisors proved to be an integral part of COSAC throughout the grant period. They assisted STEM majors in study strategies, time management, STEM tutoring and navigating the student experience at SJSU. At the beginning of each semester, COSAC peer advisors staffed desks, strategically located in College of Science buildings to answer student questions, give directions and help students with class schedules. The peer advisors also represented the College of Science at various activities, including Admitted Spartan Day and first year student orientations. These peer advisor activities have been sustained after the grant ended.

During the STEP grant period, both the Colleges of Science and Engineering implemented mandatory academic advising each semester. An advising hold was placed before the next semester's registration period and students needed to see their academic advisor to discuss their academic plans before the hold was lifted. Once the hold was lifted, students were allowed to enroll in their classes. This type of intrusive academic advising was critical and mandatory for STEM student success, especially in monitoring students' progress and pass rates in STEM gateway courses. Both colleges continue to use this type of intrusive advising strategy to monitor STEM student success and progress to degree.

Initially, there was some faculty resistance to advising changes due to the misconception of the amount of additional time required for intrusive advising, but each department worked out strategies for advising their majors. The larger departments (Biological Sciences, Chemistry and Computer Science) used a combination of staff academic advisors and faculty advisors, while the smaller departments (Geology, Mathematics, Meteorology & Climate Science, Physics & Astronomy) used faculty advisors. However, these smaller departments relied on COSAC for General Education advising and transfer articulation issues.

COSAC's advising service model has been successful, as demonstrated by increased retention rates for College of Science undergraduates as shown in Table 1. Note that for each cohort, by the end of the available observation time, an improvement of at least 10% points is observed. This model has been adopted by several academic colleges at SJSU and reflects the campus-wide influence of this project. The College of Engineering Success Center has now offered a similar advising service center for three academic years. During the time period of the STEP grant, two other colleges opened advising centers that were modeled on COSAC: the College of Applied Sciences and Arts' Student Success Center in February 2011 and the College of Social Sciences' Student Success Center in Spring 2012. Both of these centers currently have faculty directors, peer advisors and administrative staff.

Even with mandatory advising, the College of Science still had some student success issues, especially with students on university probation. We wanted to catch students earlier to prevent students from continually repeating key gateway STEM courses and ultimately being disqualified from SJSU. From examining transcripts of students who were either on probation or close to probation, we realized that we needed an earlier intervention program. Our strategy involved creating a 3-unit probation course as well as working with campus data systems to create a mechanism to pull final course grades from key gateway courses for routine analysis.

Table 1. San José State University College of Science First Time Student Retention Rates from Fall 2006 through Fall 2011

	Fall 2006 (253 students)	*Fall 2007 (317 students)*	*Fall 2008 (338 students)*	*Fall 2009 (298 students)*	*Fall 2010 (314 students)*	*Fall 2011 (314 students)*
1st Year	78.3%	79.7%	83.1%	86.6%	88.3%	88.2%
2nd Year	66.0%	72.2%	76.9%	79.9%	79.9%	
3rd Year	59.7%	65.8%	72.7%	75.8%		
4th Year	55.3%	64.6%	67.7%			

Data obtained from www.iea.sjsu.edu/reports/ssm. Rates are calculated from the original entering cohort. These retention rates are for students who remained COS majors.

In Spring 2009, SJSU did not admit any new transfer students. We also reviewed our probation students from Fall 2009 and concluded that many of our new probation students were either new first year or transfer students who entered SJSU in Fall 2009. Thus, we re-tooled our "first-year" experience course for transfer students to create a course for Science students on probation.

Today, the course also serves probation students from the Colleges of Applied Sciences and Arts, Business, Engineering and Social Sciences. In the College of Sciences, those probation students who cannot enroll in this course due to a conflicting schedule or other reasons are required to attend a probation workshop and meet with COSAC peer advisors during the semester that they are on probation. From Spring 2009 through Fall 2011, 274 STEM students participated in our probation intervention programs with an average of 70% returning to good academic standing after one semester. An average of 88% of these students improved their SJSU cumulative GPA after participating in either intervention. From Spring 2009 through Spring 2012, approximately 65% of probation STEM majors were retained as STEM majors through either intervention. Unfortunately, comparative data prior to this intervention is not available.

As part of our STEP grant, we partnered with SJSU's Career Center because we observed that many entering students really did not understand what STEM careers offered them. Funding from this grant provided resources for a Career Center graduate student intern who was assigned to work with College of Science majors. One of the first deliverables was Science Exploration Sheets that provided students with career information for our majors. This intern participated in the existing College of Science's First Year Experience Course that is required for all new College of Science majors who are classified as remedial. Over the course of the grant, the Career Center had almost 1500 science students registered at the Career Center as new and active registrants. Moreover, the Career Center noted an increase in STEM students visiting the Career Center who were on academic probation. College of Science students also utilized the Career Center for assistance with preparing resumés and with choosing/changing majors.

Under the STEP grant, we established a data infrastructure between the College of Science and our Office of Institutional Effectiveness and Analytics to accurately and comprehensively track STEM student progress in gateway coursework, annual retention rates and graduation rates. The overall sustained outcome was to create a university-wide student success electronic milestone dashboard (www.iea.sjsu.edu/reports/ssm). This dashboard allows anyone to monitor undergraduate first year or transfer student cohort progress using the following milestones: orientation, remediation, coursework, general education bottleneck courses, retention, upper division writing skills test and course, and graduation. These specific milestones were chosen based on student success research conducted by Institution for Higher Education Leadership and the Education Trust (*11*).

Providing Professional Development Opportunities for Faculty Who Teach STEM Gateway Courses

Professional development opportunities were provided for faculty in the Departments of Mathematics, Computer Science and Physics & Astronomy in order to gain new knowledge about high-impact pedagogical approaches and course revision ideas. The opportunities were offered through a department-specific approach in which department faculty identified specific professional conferences, workshops and experts to consult in order to seek out best practices. Both Mathematics and Physics & Astronomy chose to adapt Peer-Led Team Learning (PLTL) into their STEM gateway courses (*3*). It should be noted that Chemistry has had an extensive PLTL-like program in both their general chemistry and organic chemistry courses for many years.

For many years, the math department struggled with lower than acceptable passing rates in pre-calculus and calculus (approximately 60-65%). Both the Colleges of Science and Engineering determined that a significant number of science and engineering majors were repeating pre-calculus and calculus several times to achieve a grade of C- or better before advancing to the next course. Through this grant, the math department was able to research "best practices" in supplemental instruction (SI) and adapted the math supplemental program, based on PLTL, from California State University Los Angeles. These SI workshops have now been institutionalized in the following courses (including both STEM and non-STEM majors): College Algebra, Pre-calculus, Calculus I/II/III (STEM majors) and Business/Aviation Major Calculus. Passing rates in these courses have dramatically increased (up to 75%) with the addition of the supplemental workshops. These workshops have led to increased retention for STEM majors and have been sustained even under budget cuts.

For the physical classrooms in which these workshops occur, the perimeter of the walls is covered with white boards to encourage students to get out of their desks and to work together to solve problems in mathematics. With a predominantly commuter campus, having students remain on campus in a supplemental structured learning environment also contributes to their success.

Supplemental workshops in our calculus-based physics classes were added, but the results did not show as dramatic an increase in pass rates as in our Calculus courses. The cumulative average passing rate for the first semester calculus-based physics course for STEM majors was 79.3% and 84.7% for the second semester calculus-based physics course (Fall 2008 through Spring 2012). Before this intervention, the cumulative average passing rate for the first semester course was approximately 75%. Data is not available for the second semester. However, prior to these workshops, student aptitude was at 15% based on the "Force Concepts Inventory (*12*)" and increased to 26% with the introduction of the supplemental workshops. These workshops have not survived budget cuts, so the department is looking at additional less costly student-centered strategies for these courses.

Computer Science faculty examined pedagogies to better reach their students in their introductory computer science gateway courses in which the pass rate was low. For example, faculty noted that from Fall 2005 through Spring 2009, students in the introductory computer science course had a high failure rate (42% D/F/W). In response, through this STEP grant, several Computer Science faculty took part in professional development opportunities (workshops, immersion in computer science pedagogy literature, sharing of best practices) and engaged in the following activities:

1. Faculty switched from traditional lectures to short lectures plus active learning labs.
2. Students did pre-class readings and pre-class quizzes through an on-line learning management system.
3. Students kept on task with a two-stage delivery of homework (draft, then final version).
4. Students used laptops for exams.
5. Computer Science peer mentors met with students to coach them and provide social integration (accompanied mentees to department functions and introduce them to the Computer Science Club).

With these interventions, the "D/F/W" rate has decreased on average to 31% from Fall 2009 through Spring 2011.

These results indicate that faculty members do indeed respond from researching best practices for student-centered learning in their disciplines. They continue to adapt their lesson plans to obtain course learning outcomes with increased passing rates through a student-centered, research-based approach.

Immersing STEM Majors into Comprehensive Learning Communities

For the College of Science, our First-Year Experience course continues to provide our new College of Science first year students with small activity sessions in which students work together in a learning communities (*13–16*). This course is mandatory for all remedial first year College of Science students. These students are encouraged to keep their learning community as they move forward in their STEM coursework. Table 2 displays the retention rates for these students, again noting an increasing trend in retention rates through the STEP grant period. These activity sessions are continuing even though STEP funding has ended. Throughout the grant, two major social events such as the College of Science First-Year Experience Thanksgiving Lunch and the Spring BBQ were held and established as STEM traditions at SJSU. These events brought together students, faculty and staff from previous cohorts of College of Science First-Year Experience courses. These events continue to be consistently well attended (300-400 student majors) and serve as important community building gatherings for STEM majors, creating a sense of belonging to the College of Science and San José State University.

Table 2. Retention Rates from Fall 2006 through Fall 2011 for San José State University College of Science (COS) First-Time Students Enrolled in the COS First-Year Experience Course

	Fall 2006 (112 students)	*Fall 2007 (159 students)*	*Fall 2008 (217 students)*	*Fall 2009 (140 students)*	*Fall 2010 (144 students)*	*Fall 2011 (103 students)*
1st Year	78.6%	81.1%	82.5%	90.0%	89.4%	88.4%
2nd Year	67.0%	73.6%	75.6%	80.7%	80.9%	
3rd Year	56.2%	68.6%	72.4%	78.6%		
4th Year	50.9%	66.0%	67.3%			

Data obtained from www.iea.sjsu.edu/reports/ssm. Rates are calculated from the original COS First-Year Experience cohort. These retention rates are for students who remained COS majors.

The College of Science also identified study rooms within each department where STEM students can work together on problem sets, study and/or receive tutoring assistance out of class. Many of these rooms were repurposed from the department student organizations that already had meeting rooms. These spaces also facilitated social integration that is key to student success and matriculation.

After the STEP Grant—Sustaining the Effort

With shrinking state resources, it has been a challenge to sustain the efforts initiated by the STEP grant project. Fortunately, SJSU university administrators recognized the importance of our STEP grant, especially the critical role of the COSAC in supporting student success. Coincidentally, an opportunity that arose during this grant was the California State University's call to develop a comprehensive campus-wide strategy to improve our retention and graduate rates. This mandate provided us with the opportunity to expand our successful efforts in our STEP grant throughout campus. With financial support now from a special Student Success and Technology Excellence fee, other advising and student success centers are ensured funding for staff academic advisors, an administrative assistant and five peer advisors. The supplemental instruction courses are still a challenge to offer, but our calculus and chemistry workshops continue.

It is also important to remind chemistry faculty, especially those faculty who teach gateway courses, to be vigilant about campus student success efforts. These instructors represent a type of "first responder" for students and can help them successfully transition to the university learning environment. If a four-year institution has a large number of community college transfer students, our experience with STEP suggests that faculty, chairs, deans and advisors should meet at least once a year to provide updates on enrollment and transfer student success. Another concern is that it is not uncommon for STEM transfer student populations to fail to complete all of their lower division STEM major courses at their two-year colleges. In the best interest of STEM student success, this concern needs to be discussed with among two-year and four-year college faculty and administrators. Organization of meetings between two-year and four-year college partners (mostly with the advisors and articulation officers) allows for important communication avenues. This enhanced communication can be used to remind those who work with two-year students that they should complete all general education and lower division major requirements before transferring.

The key to increasing retention and graduation rates includes more than just the STEM curricula. Students need guidance and mentoring to be successful STEM students. They also need to feel that they are part of the STEM community and understand STEM pathways to careers and post-baccalaureate education opportunities. Institutions should provide tools to track student success milestones to allow careful monitoring of students' progress in key courses. Assessment strategies should be used to refine curricula. Academic advising needs to be fully supported by everyone so STEM students understand their degree pathways. If STEM students find themselves on probation, the department and college should provide proactive interventions to assist students to get back on track. These are some of the lessons learned from our STEP grant.

In conclusion, SJSU's STEP grant catalyzed a change of culture that led to a focus on students and their success as STEM majors. We are happy to note that our STEP efforts for the most part have been sustained within the College of Science and have been adopted by other colleges at SJSU.

Acknowledgments

I wish to thank Dr. Susan Hixson for her support for this project and especially the vision of expanding my work from the "New Traditions" project to this project and beyond. I also wish to sincerely thank the following individuals for their efforts on our STEP grant: Dr. Dan Walker (PI), Ms. Ann Baldwin (grant administrative assistant), Dr. Rona Halualani (evaluator), Mr. Michael Randle (academic advisor and SCI 2/90T instructor) as well as Mr. John Kim and Ms. Ana Drajovic (founding COSAC lead peer advisors). This project was supported by a grant from the National Science Foundation STEM Talent Expansion Program Type 1A, Grant #0653260.

References

1. Barrow, G. M. *J. Chem. Ed.* **1999**, *76* (2), 158.
2. Landis, C. R.; Peace, G. E., Jr.; Scharberg, M. A.; Branz, S.; Spencer, J. N.; Ricci, R. W.; Zumdahl, S. A.; Shaw, D. *J. Chem. Ed.* **1998**, *78* (6), 741.
3. Gosser, D. K.; Cracolice, M. S.; Kampmeier, J. A.; Roth, V.; Strozak, V. S.; Varma-Nelson, P. *Peer-Led Team Learning: A Guidebook*; Prentice Hall: New Jersey, 2001
4. Astin, A. W.; Green, K. C.; Korn, W. S. *The American Freshman: Twenty-Year Trends, 1966–1985*; Higher Education Research Institute, UCLA: Los Angeles, CA, 1987.
5. Astin, A. W.; Astin, H. S. *Undergraduate Science Education: the Impact of Different College Environments on the Educational Pipeline in the Sciences*; Higher Education research Institute, UCLA: Los Angeles, CA, 1993.
6. Pascarella, E. T.; Terenzini, P. T. Predicting voluntary freshmen year persistence/withdrawal behavior in a residential university: A path analytic validation of Tinto's model. *J. Educ. Psychol.* **1983**, *75* (2), 215–226.
7. Seymour, E.; Hewitt, N. M *Talking About Leaving: Why Undergraduates Leave the Sciences*; Westview Press, a division of Harper Collins Publishers: Boulder, CO, 1997.
8. Tinto, V. *Leaving College: Rethinking the Causes and Cures of Student Attrition*, 2nd ed.; University of Chicago Press: Chicago, 1993.
9. Tinto, V. *Completing College: Rethinking Institutional Action*; University of Chicago Press: Chicago, 2012.
10. Braxton, J. M.; Sullivan, A. S.; Johnson, R. M. Appraising Tinto's Theory of College Student Departure. In *Higher Education: Handbook of Theory and Research*; Smart, J. C., Ed.; Agathon Press: New York, 1997; Vol. 12.
11. Offenstein, J., Moore, C.; Shullock, N. *Advancing By Degrees: A Framework for Increasing College Completion*; Institute for Higher Education Leadership and The Education Trust: Sacramento, CA, 2010. http://www.csus.edu/ihelp/PDFs/R_AdvbyDegrees_0510.pdf.
12. Hestenes, D.; Wells, M.; Swackhamer, G. Force concept inventory. *Phys. Teach.* **1992**, *30*, 141–158.
13. Fredericksen, E. Minority Students and the Learning Community Experience: A Cluster Experiment. *Paper Presented at the Annual Meeting of the Conference on College Composition and Communication*, Chicago, IL, April 1–4, 1998.
14. Lenning, O. T.; Ebbers, L. H. The Powerful Potential of Learning Communities: Improving Education for the Future. *ASHE-ERIC Higher Education Report*; National Academy Press: Washington, DC, 1999; Vol. 26, No. 6.
15. Poisel, M. A.; Ehasz, M. *University of Central Florida: Energizing a Team to Support Seamless Transition*, 2003. http://www.sc.edu/fye/events/presentation/sit/2003/pdf/adr.pdf (accessed December 17, 2003).
16. Stuart, M. H.; Cowans, R.; Heath, M. *Policy Center on the First Year of College. Campus Practices Serving To Facilitate the Transfer Transition*, 2000.

Editors' Biographies

Thomas Holme

Thomas Holme is a Professor of Chemistry at Iowa State University and the Director of the American Chemical Society (ACS) Examinations Institute. He received his B.S. in Chemistry and Physics from Loras College and his Ph.D. in Chemistry from Rice University. His research group conducts studies in both computational chemistry and in chemistry education research. In chemistry education, the emphasis is on characterizing measures of student learning. He is a Fellow of the ACS and of AAAS. He has been a Fullbright Scholar and the recipient of the Helen Free Award for Chemistry Outreach from ACS.

Melanie M. Cooper

Melanie M. Cooper is the Lappan-Phillips Professor of Science Education and Professor of Chemistry at Michigan State University. She received her B.S., M.S., and Ph.D. in chemistry from the University of Manchester, England. Her research has focused on improving teaching and learning in large enrollment general and organic chemistry courses, and she is a proponent of evidence-based curriculum reform. She was a member of the inaugural class of ACS Fellows and is also a Fellow of the AAAS. She has received a number of teaching awards including the 2010-2011 Outstanding Undergraduate Science Teacher Award from the Society for College Science Teaching and the 2013 James Flack Norris Award.

Pratibha Varma-Nelson

Pratibha Varma-Nelson is Professor of Chemistry and the Executive Director of the Center for Teaching and Learning at Indiana University-Purdue University Indianapolis. She received her B.S. in Chemistry from University of Pune, India, and M.S. and Ph.D. from the University of Illinois in Chicago. She was a member of the leadership team that worked on development, implementation and dissemination of the Peer-Led Team Learning (PLTL) model of teaching. Currently her group is working on research and development of cyber-PLTL (cPLTL). Her teaching awards include 2008 James Flack Norris, 2011 Stanley C. Israel Award from ACS, and in 2012 Sloan-C Award for Effective Practices in Online and Blended Education.

© 2013 American Chemical Society

Indexes

Author Index

Subject Index

I

L

M

N

P

R

U

W